After Dolly

After Dolly

The Uses and Misuses of Human Cloning

IAN WILMUT and ROGER HIGHFIELD

LITTLE, BROWN

LITTLE, BROWN

First published in Great Britain in 2006 by Little, Brown

A CIP catalogue record for this book
is available from the British Library.

Hardback ISBN-13: 978-0-316-73320-5
Hardback ISBN-10: 0-316-73320-2
C Format ISBN-13: 978-0-316-72469-2
C Format ISBN-10: 0-316-72469-6

Typeset in Bembo by M Rules
Printed and bound in Great Britain by
Clays Ltd, St Ives plc

Little, Brown Book Group
Brettenham House
Lancaster Place
London WC2E 7EN

A member of the Hachette Livre Group of Companies

www.littlebrown.co.uk

To the tens of millions of people
who will one day benefit from research on
cloning, embryos and stem cells.

Contents

List of illustrations

Dolly with her surrogate mother soon after birth.
Roslin Institute seen from the air.
Dolly with three of her lambs, all born naturally.
Dolly with Harry Griffin.
The celebration when Megan was ten years old in July 2005.
Bill Ritchie seen at his microscope with manipulator equipment.
Partway through the process of enucleation.
The chamber on the microscope in which the eggs are held for manipulation.
To confirm enucleation the pipette is held in UV light.
The much smaller nuclear donor cell is picked up by the pipette.
Electric pulses are used to fuse the two cells together.
Sometimes we assess development of the eggs by staining them to reveal the nuclei, seen here as the small bright bodies.
The development of a foetus can be monitored by use of ultrasound scans.
Dolly at maturity seen in the paddock outside her barn.
The Roslin nuclear transfer group.

Acknowledgements

We owe a great debt of gratitude to many people. Very many thanks to John Brockman and Katinka Matson, who provided the encouragement for the two of us to collaborate. Thanks are also due to Angela von der Lippe and Vanessa Levine-Smith of Norton in America. For the British edition, we are grateful to Tim Whiting and Iain Hunt of Little, Brown, and to John Woodruff. Neither of us could function in the office without the help of two people: Gulshan Chunara at the *Daily Telegraph* and, at the Roslin Institute, Lynne Elvin. Thanks are also due to the former editor of the *Telegraph*, Martin Newland, for allowing Roger to pursue this project.

The following people have answered questions, sent papers, and provided other invaluable advice. Some of them have also commented on drafts or chapter sections (any remaining howlers are, of course, ours alone). So, many thanks to: Michael Banner, Derek Bromhall, Keith Campbell, Robin Christmas, Alan Colman, Joy Delhanty, Chris Evans, Matthew Freeman, Chris Graham, John Gurdon, John Harris, Doris Highfield, Tim King, Robin Lovell-Badge, Dusty Miller, Björn Oback, Jean-Paul Renard, Susan Rhind, Bill Ritchie, Jim Robl, Linda Rosenbaum, Jane Skok, Davor Solter, Miodrag Stojkovic, Andrzej Tarkowski, Ted Tuddenham, Karen Walker and Steen Willadsen. Photographs were provided by Norrie Russell of the Roslin Institute.

Graham Farmelo and Paul Carter gave us crucial feedback on

how to structure chapters to make the book more accessible. Eamonn Matthews provided transcripts of interviews of his award-winning series on IVF, *Baby Makers*. And Lois Godfrey kindly gave us permission to quote her uncle, J. B. S. Haldane.

Finally, heartfelt thanks are due to those closest to us for tolerating our need to invest what little spare time we have in this project. Roger would like to thank his wonderful wife, Julia, for all her support and love, his children Holly and Rory for lifting his spirits, and their au pair Jeanett Svenkerud, grandparents Jim and Betty Brookes, and the Dads' Club (Simon Singh and Raj Persaud) for helping to keep them amused. Ian is grateful to his wife, Vivienne, for her continued tolerance of the time that he puts into his work and for her infinite patience and support.

Roger Highfield and Ian Wilmut
July 2006

Chapter 1

A Star Is Born

Few celebrities have so captivated the world's media as she did. Her arrival caused a sensation, triggering an orgy of speculation, gossip and hype. She posed for *People* magazine, became a cover girl and even caught the eye of Bill Clinton. Plays, cartoons and operas were inspired by her story, advertisers traded on her image, and her name seemed to be on everyone's lips. Those closest to her said that the attention rather went to her head, and she soon started putting on airs.

Nor was her career slowed by motherhood; even when she had six young ones to care for, she continued to be in the news. But, as with so many global superstars who live life in the fast lane, things were destined to go wrong. She developed a cough and began to go into decline. After a few weeks, she died tragically young, of lung cancer – even though she had never smoked. Her passing marked the end of a great celebrity life. But this was no ordinary diva. This was a sheep.

So much for the breathless media account of the life and times of Dolly the sheep. Her story is by now as much a part of the history books as of the annals of science. Her birth at the Roslin Institute, near Edinburgh, on 5 July 1996 heralded a new era of biological control. Working with a large team, I became the first to reverse cellular time, the process by which embryonic cells differentiate to become two hundred or so specialised cell types in the body.

We had defied the biological understanding of the day: that in nature development runs in one direction only. Cells in tissues as different as brain, muscle, bone and skin are all derived from one small cell, the fertilised egg. Until Dolly was born, it was thought that the mechanisms that picked the relevant DNA code that made a cell adopt the identity of skin, rather than muscle, brain or whatever, were so complex and so rigidly fixed that it would not be possible to undo them. This deeply held conviction was overturned in Dolly. She was the first mammal to be cloned from an adult cell, a feat with numerous practical applications, many of which would raise profound moral and ethical issues.

The term 'clone', from the Greek for 'twig', denotes one of a group of identical entities. Dolly was (almost) genetically identical to a cell taken from a six-year-old sheep, the nucleus from which had been transplanted into an egg cell from a second sheep, and then inserted into the uterus of a third sheep, and then a fourth, to develop. It was because the process started with a mammary cell from an old ewe that she was named Dolly – our affectionate tribute to the buxom American singer Dolly Parton.

While we now take for granted that it is possible to clone adult animals, the birth of Dolly shocked members of the general public who dwelled on the implications of reproduction without the act of sex. The feat shocked many in the research community too. Scientists were often heard to declare that this or that procedure would be 'biologically impossible', but with Dolly that expression had now lost all meaning. Some pundits have even said that Dolly broke the laws of nature. But she underlined, rather than defied, those laws. She underscored how, in the twenty-first century and beyond, human ambition will be bound only by biology and by society's sense of right and wrong.

Why Clone?

When we created Dolly, we were not thinking about rooms full of clones, or creating hillock upon hillock of identical sheep to guarantee a good night's sleep. We were not thinking about helping lesbians to reproduce without the help of a sperm bank or about multiplying movie stars. We were certainly not thinking of duplicating dictators.

The tortuous tale of Dolly does not fit the storyboard of a traditional Hollywood movie: my single-minded struggle, against all odds, to fulfil my dream; strange goings-on in a subterranean laboratory one dark and stormy night; or how a tight-knit group of hirsute underdogs toiling in an obscure Scottish lab snatched the cloning prize from arrogant, clean-shaven, lantern-jawed scientists working in a well-funded North American powerhouse of genetics. True, my team did sport a fair amount of facial fuzz and long hair, but it was a special combination of factors that led to success, along with a spark of serendipity. The right people with the right skills and understanding came together at the right time and in the right place – the Roslin, with its collection of pristine molecular biology laboratories, hay-strewn barns and up-to-date facilities for surgery, whether performed on livestock or on individual living cells.

Dolly's story is all the more remarkable for other reasons. The classic scientific scenario of rival teams racing for a prize, as vividly depicted in Jim Watson's *The Double Helix*, did not apply in our case. We were spurred on mostly by pure curiosity, though we did have in mind practical applications in research and agriculture. As in any government laboratory, research was a grind because money was always in short supply – even more so in those days, when the science establishment was subjected to punishing cutbacks: on the day Dolly was unveiled, the government withdrew its funding for my project. The biotechnology company that helped us create her, PPL

Therapeutics, was our ally but not our friend, in the sense that we shared information only on a need-to-know basis.

The group was led by myself and Keith Campbell, a cell biologist ten years my junior, who had built up a profound understanding of what makes cells tick. It is still a surprise that we were so effective together because neither of us is well organised. His adventurous nature and knowledge of cells complemented my own scientific curiosity, my experience with cloning, and my strategic, step-by-step approach to achieving the goal of the genetic transformation of livestock. I conceived the project, but it was thanks to Keith's inspiration that it succeeded.

We were united by more than this, however. Both Keith and I had become disillusioned with life at the Roslin. As an embryologist I had been forced to give up my research on prenatal mortality to focus on the genetic alteration of animals – work which, ironically, would eventually give me huge satisfaction and pave the way for Dolly. Keith, as the new member of the team who had key scientific insights, found it frustrating selling his ideas to an established group. He felt resented and was unhappy – at least until our work together started to pay off.

Both of us had thought about leaving the Roslin. But we both had pragmatic reasons, to do with home and family, for remaining in what was a stunning part of Scotland. There was a castle atop an almost isolated rock to the south-south-east of Roslin village. We were surrounded by bleak, uninhabited hills, rising to almost two thousand feet and stretching some sixteen miles from the outskirts of Edinburgh to Lanarkshire. The landscape of uplands, lochs and glens was just the place to follow the traditional pursuits of drinking whisky, hill walking and staring into space. It was an ideal place to raise young children. I loved to jog in the Midlothian countryside and became reasonably adept at, and addicted to, the peculiar Scottish game of curling (if you are unfamiliar with it, think of bowls on ice).

Roslin the place, not the institute, was the attraction for us both during those difficult days. Keith and I would continue to drive to the laboratory together and have animated chats about the details of cloning. We were both excited by the science. We both hailed from the Midlands. But we were no Watson and Crick engaged in a race with rivals. We had different styles and outlooks, and we did not socialise much.

Our shared aim was to develop methods for nuclear transfer in sheep that could be used to introduce precise genetic changes. It may seem paradoxical that the wherewithal to create genetically identical offspring can also be used to induce genetic changes in animals. The answer to the riddle is that you can use cloning to make a whole animal from a cell upon which you have carried out successful genetic surgery – the one cell among the billions of others on which the surgery was botched or incomplete or had failed. That would offer huge advantages over the old-fashioned brute-force method in which we had to insert embryos into sheep time and again with no clear idea of whether the embryos had been successfully modified.

Now we had just £20,000 to work with, which a Roslin committee had offered us to do some more experiments because of the good progress we were making with cloning. Of course, because of our previous experience we would use sheep. That meant that we had to deal with their notorious obstetric problems (shepherds say that sheep spend their time dreaming up new ways to die). We were slave to their breeding cycles, which required us to work intensely during the winter, when sheep mate and conceive. The advantage, however, was that sheep were astonishingly cheap. In those days it was possible to buy one at the market for less than the cost of a bottle of mineral water in a posh hotel, and for around one five-hundredth the cost of a cow. But we were confident that, because of their similar reproduction and embryology, any method developed for sheep would also prove suitable for cattle. In effect, the sheep were small, cheap cows.

It all seemed so simple. We would use special cells from embryos – stem cells – for cloning. These grow in the laboratory, but retain many of the characteristics of embryonic cells. This was why I was optimistic that this approach to making genetic changes in animals would work. But there was a problem: we could not multiply those sheep stem cells in the laboratory (we still cannot). In our failed attempts, though, we did grow more mature cells that, as scientists say, had begun to differentiate.

Thanks to the insights of Keith Campbell into the mechanics of the cell cycle, we went on to make a remarkable find: it was possible to put these more differentiated cells into a special state (a resting state, called quiescence, which is unattainable by embryonic cells), and we discovered that we could clone them. Megan and Morag, Welsh Mountain sheep, were born as a result of this crucial insight. They were the first clones from differentiated cells. We had failed to grow stem cells, but Keith's new method of cloning had proved more powerful that we had imagined. Keith had always believed that cloning with adult cells would become possible, and with this new encouragement we set out to test the idea. The birth of Dolly confirmed the point emphatically.

As I waited for Dolly to be born during the summer of 1996, my mood would swing between elation and fear. Keith and I were confident of success, but with such a complex process involving many steps and people, failure was always a strong possibility. And if we succeeded, the thrill of it all would be tempered by the thought of the fuss that would follow. I am a private person, and I knew that my life would no longer be quiet.

And succeed we did, but only just – in one out of 277 attempts. Dolly weighed in at 6.6 kilograms, heavy for a Finn Dorset but not surprisingly so. Perhaps in another season we would have had twenty Dollys. But it was much more likely that we would have failed. Despite the emphasis on objectivity and rationality, even in science it helps to be lucky. For Keith, however, Megan and Morag were the real stars of the scientific show. He considered Dolly merely 'the gilt on the gingerbread'. We had agreed that while he would hold the coveted position of first author of the paper we would publish on Megan and Morag, I would be first author of the paper in which we described Dolly's difficult passage into the world.

We kept Dolly's existence secret for the first six months, while the paper with its details of how we had created her was scrutinised by other scientists – a routine prerequisite for publication in an academic journal. Once the news of her birth leaked out, in February 1997, she made headlines, capturing the imagination of commentators, columnists and opinion writers across the planet.

'Researchers Astounded ... Fiction Becomes True' and 'Dreaded Possibilities Are Raised' was the quaint way that one newspaper, the *New York Times*, greeted the news of Dolly's birth. But it was not the only one to sound the alarm: other headlines included 'Cloning Discovery Has Unleashed a Wolf in Sheep's Clothing', 'Cloned Sheep in Nazi Storm', 'Dolly Opens Door for Life After Death', 'The Clone Rangers Need to be Stopped' and 'Golly, Dolly! It's the Abolition of Man'. The *Weekly World News* quoted a terrified researcher who described how Dolly had killed a lamb, then eaten her victim: 'When you do something to anger her, she looks at you with those intense red eyes – eyes full of hate.'

With the benefit of hindsight, this outburst looks entirely predictable and understandable (except perhaps for the red-eyed cannibal). But I am still disappointed by the sensationalism that is so prevalent in today's journalism, especially in headlines. Although I knew that some of the possibilities raised by this form of cloning are indeed worrisome, I expected Dolly's birth to be celebrated because many more applications of cloning in the farm, clinic and hospital are wonderful. How naive I was!

Members of the American media gasped that Dolly's birth 'seemed to come out of nowhere' (she was not born in America, after all). Some scientists went further. One Nobel laureate wrote to the journals accusing us of making a silly mix-up and saying that Dolly was not a real clone at all. Others thought that Dolly was a 'clone alone', a freak accident that could never be replicated. But most were convinced that Dolly was something special, and there was talk of her creation being biology's equivalent of the splitting of the atom.

I have now become accustomed to the wild swings of the media pendulum – from the sensational prospect of cures to visions of apocalypse – and I have often had my views misrepresented to make a better headline by supposedly responsible newspapers. But when Dolly was born even I got caught up in the hype. During the initial brouhaha I found myself telling one reporter that Dolly 'might reasonably claim to be the most extraordinary creature ever born'. This was an exaggeration (but not much of one). President Clinton reacted to the 'startling' news of her birth by asking experts to review the implications; the Vatican and the European Commission swiftly followed. Within a few months came the first claims that human pregnancies were under way with cloned embryos, and similar rumours have circulated ever since, not least because of the efforts of cults, madmen and maverick scientists.

In pubs, in offices and on the streets, people could not stop talking about the idea that human cloning might soon become a reality. Newspapers asked whether it would be possible to clone the dead. Some deluded souls saw Dolly as a harbinger of a form of genetic immortality; many more were fired up by the talk of a looming ethical and political nightmare. Inevitably, Frankenstein provided a cultural reference point for the discussion. *Time* magazine began its fourteen-page special report by casting me in the lead role: 'One doesn't expect Dr. Frankenstein to show up in a wool sweater, baggy parka, soft British accent and the face of a bank clerk. But there in all banal benignity he was: Dr. Ian Wilmut.'

The cover of the German magazine *Der Spiegel* depicted a regiment of Hitlers, the implication being that cloning people would somehow facilitate a kind of science-fiction tyranny under which individual identity was lost to a nameless, faceless and oppressive mass of pseudo-humans. The cover stated: 'Wissenschaft auf dem Weg zum geklonten Menschen. Der Sündenfall.' – 'Science on the way to cloned people. The fall of man.' In private, I had mulled over the possibilities of human cloning two years earlier with Keith Campbell, as he drove me to work. That was the summer in which two of our ewes had been pregnant with Megan and Morag, our first celebrity clones. Only the month after the news of Dolly had broken did I air my thoughts in public.

In Washington, I held my breath as I prepared to address a crowded meeting of the Senate's Labor and Human Resources Committee which was being held on the eve of the first meeting of President Clinton's Bioethics Advisory Commission. I was anxious about the forthcoming grilling. As it turned out, I was indeed taken aback – but because the reception was anything but hostile. One American scientist said that I would win a Nobel prize. A senator described me as 'one of the true trailblazers in human history'. Another declared that I had 'broken the biological equivalent of the sound barrier'. A wire report concluded that 'the world's best-known shepherd' was treated by Capitol Hill lawmakers 'like a media superstar'. I did not feel like one, however, dressed in my grey suit and faded tie. Nor did I deserve to take so much credit for an effort that had depended on the hard work of so many others.

Human cloning in all its forms was, of course, the main subject of discussion. One Democrat senator likened efforts to stop cloning to the Catholic Church's attempts to silence Galileo. 'Human cloning will take place,' he said. 'And it will take place in my life-time. And I welcome it.' But I did not. I told the hearing that to clone babies would be unethical and inhumane. I would back any efforts that were made on an international basis to ban the practice. Equally, I stressed how important it was not to shut the door on life-saving applications. It was here that I presented the idea of deriving cells from cloned human embryos either to study or to treat disease. As I shall explain in these pages, I remain passionately opposed to cloning babies. I have co-authored a paper published in a scientific journal with the unusually blunt headline 'Don't Clone Humans!' But I am an equally passionate advocate of a restricted form of human cloning.

The Life of Dolly

As any visitor to my office will tell you, Dolly inspired my friends and colleagues to give me endless ovine paraphernalia – sheep clocks, sheep mugs, and the like. More importantly, she inspired a whole new generation of young people to study biology (we still get a steady stream of letters and e-mails asking for information). And she inspired many crackpots, lunatics and frauds too.

Meanwhile, the Finn Dorset sheep at the focus of the furore seemed oblivious to all the angst she had caused. While hands were wrung and brows furrowed, Dolly lived a calm and cosseted life in a barn on the Roslin Institute site with Megan and Morag. Although these two other sheep had marked the first demonstration that the developmental programming of a differentiated cell could be wiped clean so that life could start afresh, Dolly had pushed them both out of the limelight.

Unfortunately for her peers, Dolly behaved as if she were already aware of her celebrity status. She was greedy and would often grab more than her fair share of the 'cake' made of barley and molasses that we fed them. She would also assert her authority by upending her trough as soon as she finished eating. This display of dominance would culminate with her placing her forefeet on the trough, puffing out her chest, and preening. She seemed to know that she was the alpha sheep of the flock.

I still think of her as the friendly face of science, however. Whereas most sheep are shy around humans, and huddle at the back of their pen when visitors arrive, Dolly seemed to bask in their attention. She trotted to the front of her stall when she saw people coming by. The woolly prima donna would bleat excitedly and jump up, front feet on the railing, and pose for the cameras. I should emphasise that this is odd ovine behaviour: sheep do not normally stand on their rear legs. She was an incredibly personable animal and was interested and curious in the world around her, at least as far as a sheep can be. Bill Ritchie, who selected the cell used to create her, remembered how, when he shouted her name, 'you would get this deep bleat from the shed'. I suspect that her sweet and outgoing nature was shaped by Pavlovian training, because she received more than her fair share of edible bribes to encourage her to flirt with photographers or appear at the front of her pen on cue for the legions of TV crews who came to visit her.

Visitors wanted to feed her too, but we urged them not to, explaining that she had a weight problem. In truth, we had contributed the most to her expanding girth by providing edible inducements but not giving her much opportunity for exercise. Eventually, Dolly joined the world's growing number of clinically obese individuals. We put her on a diet. Even her new food regime had a celebrity ring to it. 'Hay only' read the sign hanging on her pen – not quite the 'Hay diet', but close enough.

Like all her other woolly peers, Dolly hardly looked as though she could change the course of history. But she did, of course. She met reporters, photographers and film crews from across the planet. She inspired cloning references in *The Simpsons*. She was interviewed by Fay Weldon, author of *The Cloning of Joanna May*, who later said of the experience, 'I thought she was such a beautiful creature, and so clearly had a soul, that I lost all fear of cloning.' On Dolly's birthday, people would telephone to wish her many happy returns.

Dolly was fleeced in the name of medical research (her grey-white wool ultimately went to the Science Museum as a sweater designed by a schoolgirl in a competition run by the BBC's *Clothes Show*) and she posed with one Japanese couple for wedding photographs. In fact, she was photographed as much as a Hollywood movie star, and more than any other sheep in history. As an experimental animal, one whose care was subject to strict guidelines, Dolly was banned from opening shopping malls or attending agricultural shows (despite innumerable requests) and had to turn down an invitation to mingle with fellows of the Royal Society, the world's oldest scientific academy, at their summer exhibition (a cardboard cut-out went instead). Even though the Royal Society, near Pall Mall in London, has long been a favourite haunt of scientific superstars, from Newton to Crick, Dolly was not going to stroll into its marbled entrance hall. For her, the journey to London was all risk and no benefit.

Even though Dolly was confined to the Roslin, she helped to introduce me to royalty. Thanks to Dolly, I have met the Queen twice, received a medal from Princess Anne, and been reprimanded by the Duke of York when my institute struck a deal with an American company, rather than a British one. In the Academy of Achievement held at Jackson Lake in the Rockies, an event which aims to inspire young people, I met Tom Clancy, author of high-octane thrillers, and George Lucas of *Star Wars* fame, and watched as the artist Dale Chihuly created his vibrant glass sculptures. Also in attendance that evening were the great and the good, from Colin Powell and Dick Cheney to computer wizards, astronauts and Nobel laureates. Over the years I have also spoken with many politicians, from the former Father of the House of Commons, Tam Dalyell, to Edward Kennedy (I remember his big, firm handshake). And, of course, I have encountered swarms of reporters, photographers, and anchormen and anchorwomen from around the world. All of this I owe to Dolly.

By the standards of her fellow sheep on the farm, many of which were slaughtered as early as nine months, Dolly lived a very long and full life. Some scientists feared, though for no good reason, that she was sterile, but she defied her critics and went on to breed with a Welsh Mountain ram called David and have a daughter, in April 1998. Her first-born was named Bonnie because, as her vet Tim King pronounced at the birth, 'She is a bonnie wee lamb.' Dolly had two more lambs in 1999. She gave birth to a total of six lambs, all conceived naturally and all born healthy. She was the living, bleating and woolly proof that new life – in the full sense of being able to reproduce – could come from a cloned adult cell.

Dolly became an unlikely icon for the promise and the threat of biotechnology. She was featured in a 'digital video opera' by the American composer Steve Reich, making its debut, as one headline put it, at the 'Baa-rbican'. The ethical dimensions of cloning that she raised were explored in Caryl Churchill's thought-provoking play *A Number*. Appropriately enough, Scottish anti-monarchists elected her as their preferred queen.

Zanussi used her image in an advertisement for washing machines with the slogan 'The misappliance of science' (we subsequently applied for trademark protection of Dolly's image to put a stop to such depressingly downbeat campaigns). Many wanted to oblige. In late 1997, Richard Seed, an elderly Chicago physicist – not even a physiologist – announced his intention of cloning a human being within two years. Soon after, members of a religious cult based in Canada, the Raelians (followers of Claude Vorilhon, a French-born racing car driver and journalist, who says that he was given the name Rael by extraterrestrials 1.2 metres tall), revealed a grandiose vision of achieving immortality by human cloning and founded the company Clonaid to carry out its mission. Raelians believe that humankind started with the cloning of aliens 25,000 years ago, are advocates of free love (condoms mandatory) and believe that a super-being called Elohim will return to Earth in 2025 to liberate believers.

Other would-be cloners could not be so easily dismissed. In March 2001, at a press conference in Rome, an Italian and an American announced plans to undertake human reproductive cloning. The Italian member of the team was Severino Antinori, a gynaecologist notorious for having used donor eggs and *in vitro* fertilisation (IVF) to make a sixty-two-year-old woman pregnant in 1994. The American was Panayiotis Zavos, a reproductive physiologist and an IVF expert at the Andrology Institute of America, in Lexington, Kentucky. Both Antinori and Zavos glossed over the technicalities and the many concerns about the effects of cloning.

Given the unusual way she came into being, Dolly's every bleat and every baa were by then being analysed for their biological significance and the merest hint of decrepitude. She showed no real signs of being any different from the rest of the flock, however. But there were limits to what we could do to check that she was really 'normal'. For one thing, we could not scan for aberrations of ovine mind or mood that could be linked with cloning. No one knows the mind of a normal sheep, after all.

After five years free from health problems came the first sign of trouble. Dolly began to limp, eventually becoming lame in her left hind leg. She was sent a couple of miles away, to Edinburgh University's veterinary school, where X-rays confirmed that she had arthritis in her left knee. The condition was treated effectively by injections so that she had no discomfort. Nonetheless, my secretary was bombarded with calls from companies offering arthritis cures, pleading with us to try them out on Dolly.

Although arthritis is a fairly common ailment for a middle-aged sheep, it is less common to find it in the knee (the elbow is the joint that is usually affected). Perhaps, however, it was inevitable that this would develop in a corpulent sheep who had been indulged all her life and liked to stand up and beg on her rear legs. But her limp did raise the first serious questions about the safety of cloning and whether it would lead to premature ageing.

The Death of Dolly

After enjoying many years as the world's number one celebrity sheep, Dolly developed breathing problems and a cough in February 2003. We suspected that she was suffering from pulmonary adenomatosis: a disease, not uncommon in adult sheep, in which a progressive lung tumour causes emaciation, weight loss and shortness of breath. There is no effective treatment, and within a few months infected sheep often succumb to the combined effects of the slowly growing tumour and pneumonia.

We were reasonably confident of the diagnosis because the disease had already struck four of the sixteen other sheep that had been in the same barn. The victims included one of Dolly's lambs and Morag (her barn-mate and twin sister, Megan, was unaffected, underlining that environment is as important as genetics). The disease is passed on by droplets in the breath of an infected individual. Unfortunately, by the time the first sheep succumbed to the illness it was too late to stop the spread of infection, since the disease takes up to three years to incubate. All we could do was prevent any more sheep from entering the barn.

Sheep kept in indoor pens rather than open pasture are more prone to catch this kind of lung disease. Perhaps if Dolly had been allowed to roam the green fields, gulp in fresh air and gaze at the Pentland Hills, she would not have succumbed. But because of her celebrity status she had to be protected from lunatics who might want to harm her, criminals who wanted to kidnap her and pranks by local students.

To confirm our diagnosis, Tim King decided to send Dolly for a CT scan – where a cross-section of the body is taken with X-rays – at the Scottish College of Agriculture (a highly unusual procedure for a sheep, but she was special, after all). That same afternoon I was attending a meeting on the first floor of the Roslin Institute. The discussion ended when my secretary, Lynne Elvin, came in to tell

me that Tim wanted to talk to me urgently. We went downstairs to
my office, and I took his call. Dolly's infection was worse than we
had thought. The CT scan had confirmed the diagnosis and
revealed the full extent of her lung tumours. Dolly was also struggling to recover from the general anaesthetic used to keep her still
for the scan. Before her death, which now looked inevitable, Dolly
would face weeks of pain and discomfort.

For a moment there, in my office, I held my head in my hands.
Things were much bleaker than I had expected, and I was both
taken aback and genuinely sad. We'd all grown fond of Dolly and
her little quirks, and the prospect of losing her in such unfortunate
circumstances was a jolt. Nevertheless, there seemed only one
humane response. At around 3.30 p.m. on that day, Valentine's Day,
I decided that her suffering should end. Tim King put her down
with a lethal injection of barbiturates before she could come out of
the general anaesthetic. 'Dolly was almost a pet among the staff,'
King recalls. 'There were many very upset people that afternoon.
Tears were shed.'

Years before her death, when she was only two, the Royal
Museum of Scotland had approached us to make arrangements to
preserve her in the event of her death. Dolly was skinned by taxi-
dermists from the museum on the afternoon that she died, before
any decay could begin. Her skin was pickled and tanned to preserve
it, then her hide was stretched over a fibreglass mould of her body,
and glass eyes were later inserted. Her nose and other muscles were
fashioned from Plasticine to give her shape, and her stuffed body
was eventually put on display at the Royal Museum of Scotland,
mounted on a straw-covered plinth.

Immediately after she was skinned, Dolly's remains were sent for
a thorough post-mortem examination. It was carried out by Susan
Rhind, a senior lecturer in veterinary pathology at the University of
Edinburgh who has done much to document the potential harmful effects of cloning, mostly on lambs. Susan and her assistant did a

thorough job, conducting exhaustive tests and taking more tissue samples than usual, given the historic importance of their subject. Even though there was scientific interest in the viability of clones of clones, we had decided not to take samples of tissue for that purpose, and simply concentrated on assessing Dolly's physical state.

Susan confirmed that Dolly had pulmonary adenomatosis. Dolly's lungs contained large areas of firm, grey solid tumour. Around the central mass the cancer had spread to seed smaller nodules and masses into the surrounding lung tissue. There were also signs of pleurisy – her lungs were attached to the wall of her chest in places. Her larger airways and trachea contained white, frothy fluid. 'It was a classic case,' Susan said. The disease is caused by a virus, called the jaagsiekte sheep retrovirus (also known as pulmonary adenomatosis virus). The virus infects lung cells responsible for secreting lung fluid (alveolar type II cells); by transforming these cells and thereby expanding their numbers, it boosts the secretion of lung fluid and the production of viruses. Cancer results because the envelope of the virus contains a protein that makes infected cells divide uncontrollably.

Susan also revealed the extent of the arthritis in Dolly's hind legs. The knee – the stifle joint – of her left leg was the worst affected of all. Concern over potential harmful effects of cloning was now being voiced with increasing frequency, both in academic circles and in the media. Perhaps Dolly, being the first clone of an adult cell, would reveal all kinds of defects. But in fact the post-mortem had revealed nothing particularly unusual for an animal of her age and weight.

Because I had been so keen to end Dolly's suffering, I had given no thought to dealing with the media. We thought at first that we had better keep her sad passing to ourselves over the weekend so that we would have time to inform the acting director, Harry Griffin, who was away from the office, and prepare a full press release. But so many people had become involved with Dolly's last day in one way or another that the news was bound to get out. So we decided to make Dolly's death public immediately, for any hint of secrecy would have fuelled the fertile imagination of journalists, who would have speculated on what it was about her death that made us want to cover it up. We did our best to make the facts known. Even so, her death was often wrongly assigned to premature ageing caused by cloning. E-mails and cards followed, from scientists and public alike, saying how sorry they were that she had gone. The phone did not stop ringing for a week or so.

Dolly was survived by five of her lambs, and I am sure they remembered her. Work by Keith Kendrick at the Babraham Institute, Cambridge, suggests that sheep have a much richer inner social and emotional life than we have given them credit for. They can remember at least fifty sheep faces for many years, even in pro-file, when most humans would be pushed to tell the animals apart. These animals can also remember ten or more familiar human faces, and their brains, like ours, seem capable of forming mental images.

So while apparently ruminating mindlessly, Dolly and her kind could be dwelling on long-absent flock-mates, mothers, or even a significant human in their life, like me. Even in her woozy state after her anaesthetic, Dolly may even have been able to detect the dis-comfort of Tim King when he ended her life. Sheep, like a number of other animals, are better than people at detecting bodily changes that occur during emotional or psychological states in others, even if they cannot imagine what that other creature is actually thinking.

The Monday after her death, on 17 February, I held a wake at the Roslin to mark Dolly's passing, to signify the importance of the experiment and to say thanks to all the people who had looked after her. With champagne, we toasted Dolly to celebrate her extraordi-nary life. The carers, vets and scientists had a chance to bid farewell to this remarkable animal. Her passing was a sad time for those, like me, who had been present when she began her life in a Petri dish, visited her often in her stall, shown her off to the world and discussed her almost every day of her life. Even today, when I pass by the barn where Dolly lived, I am reminded of her.

Many media commentators concluded that cloning was to blame for her death. Typically, sheep can live for eleven or twelve years, yet Dolly managed only six. One newspaper announced that 'Cloning may have consequences far beyond a creature's birth.' Another stated, with grim relish, 'Dolly the sheep may have been fated to die prematurely from the moment she was conceived.' She did indeed seem to have inherited six years of wear and tear from the udder cell used to clone her: studies by Wei Cui, one of my colleagues, found evidence from structures in cells called telomeres that Dolly's biology was older than her years – her telomeres were shorter than expected and more typical of an older sheep. Was it that, like most copies, she had faded a little quicker than the original?

Perhaps. But you cannot infer much from a single case, particularly when there were no sheep to compare her with: we do not know enough about the typical illnesses that affect a six-year-old sheep. In fact, no one could even state the 'normal' expectations of ageing or her average life expectancy with confidence, since we tend to eat most sheep by the time they are about nine months old. From the example of Megan, one of the clones in Dolly's shadow, Dolly did seem to die young – I celebrated Megan's tenth birthday on 19 June 2005, with members of the original team who helped to create her.

But Dolly's progressive lung disease, which also claimed the life of Megan's twin, probably said more about being kept in a barn than about being cloned. To confuse matters more, there is also individual variation in any animal species. Other studies, to which I shall return, say much more about the dangers of cloning, from super-sized offspring and breathing difficulties to problems with immune systems and blighted pregnancies. Dolly's post-mortem revealed a specimen that was unremarkable for a middle-aged sheep with a weight problem who had enjoyed a celebrity lifestyle. There was

nothing exotic about the viral disease that prompted her euthanasia. Dolly was unscathed by cloning, given that she was a pioneer of
the technique.

After Dolly

Near the Roslin's Large Animal Unit, a few yards from a public path
where thousands of people stroll each year, is an unremarkable
building that is still called the Visitors' Centre. But this is no pur-
pose-built facility offering coffee, posters and souvenirs. It is simply
a barn next to a small field, and was so named because of the hundreds of journalists, politicians, dignitaries, and others who had
made the pilgrimage there to meet Dolly. Years after her death, the
name has stuck, a tribute to the star qualities of this animal.

Throw open the doors today and you will see a little flock of five
clones in a straw-lined pen, the only living evidence of the years of
work by our large team: Megan; two sheep from cloning experiments in which an attempt was made to implant a human
blood-clotting factor; Dolly's daughter, Bonnie; and Morag's offspring, Katy. Since those heady, pioneering days, the focus of the
effort to clone farm animals has moved abroad. The team that created her has split up. Keith Campbell moved to PPL and then to
Nottingham, where he is now trying to improve the process of
cloning. PPL is, in effect, no more. And I now work for Edinburgh
University. The Roslin, with which I still collaborate, is focusing its
efforts on farm animal breeding and health.

The life of Dolly has opened up thrilling new opportunities in
medicine, but at the same time has raised serious ethical issues and
stirred many concerns. And she continues to do so. Even as I write
this, I am feeling the aftershocks of her birth. I have in the space of
a few days been criticised by an American presidential ethics adviser
for being 'misguided'. As he was declaring that genetically

engineered humans would never happen, he raised the ludicrous possibility of designing people with wings (was he thinking of angels, perhaps?). I have also been picketed by protesters in sheep masks and quoted in a tabloid article headlined 'I'm Not God'.

I believe in the right to protest. I believe equally that right-wing religious paranoia is slowing the quest for treatments, and as a result will harm people and cause suffering. I find it a constant cause of frustration that, in all the public debate, the harm that can be done to future generations by neglecting a useful technology is rarely taken into account. Disregarding the hype, of which there is much, I have no doubt about the long-term potential. The creation of cloned embryos from patients will, over the next few decades, lead to treatments for degenerative diseases such as heart disease, spinal cord injury, liver damage, diabetes, Parkinson's, motor neuron disease and Alzheimer's, all of which cause damage to cells which, subsequently, are not repaired or replaced. By creating a cloned embryo of a patient, we can obtain a source of the patient's own cells – stem cells – which can be used to understand the disease, test treatments, and not only repair a body but regenerate it too. Those cells can metamorphose into any cell type.

We cannot know yet which of these serious diseases will be treatable with cells from cloned embryos or when treatment is likely to begin, but I believe that the drive to perfect them is strong because there is no fully effective treatment for any of them, and for some there is no treatment at all. One day doctors will be able to use cloning to grow a patient's own cells and tissues so that they can carry out repairs. Cells from these embryos will also speed the search for the next generation of blockbuster medicines and help to reduce our dependence on animal research. As a bonus, this work will give us profound insights into human development and how it can go wrong, and into how to intervene at the embryo stage to prevent many terrible genetic diseases from developing.

The potential of cloning to alleviate suffering – even end it for some diseases – is so great in the medium term that I believe it would be immoral not to clone human embryos for this purpose. In the long term, a vast range of alternative and embryo-free ways to grow cells and tissues, perhaps even organs, may also rest on the foundations of this research. Cloning may by then have become viewed as a mere diversion and superfluous.

However, I want to go even further than this and propose that scientists may one day grow cloned human embryos to term to prevent the suffering caused by hereditary disease. Doctors should be able to offer at-risk couples the opportunity to conceive with IVF methods, break down the resulting embryos into cells, correct any serious genetic defects in these cells, and then clone demonstrably healthy cells to create a new embryo that can be implanted to start a pregnancy. This does add up to a qualified proposal to clone and genetically alter human beings, but I hope to build a case to show that this approach could eventually be justified because it does not, paradoxically, entail cloning a person (in the sense in which most people understand the phrase) and because it could do much to combat serious genetic disease and reduce human suffering.

My vision raises many technical questions. Scientists still have much to discover about the process of cloning and the detailed molecular mechanisms of how it goes wrong and why it does so often. Only then will they find ways to make the process of cloning safe. Only then will they have complete confidence in growing a patient's cells by the creation of embryo clones. And only then will they be in a position to weigh up my proposal that for certain diseases we should consider the cloning of IVF embryos.

My vision also faces stiff opposition. Many people will fight *any* proposal to create embryos. Many will object to *any* use of embryos in research. Many will object to *any* proposal to alter the genes of human beings. And among those with less extreme views, many are likely to feel profound unease at what I suggest. But I want to say what I think, rather than what people would like to hear, even though I suspect that my peers will consider me at best brave, at worst foolhardy and naive. Until now it has been difficult to discuss these ideas in public. The drizzle of outlandish remarks and claims from the mavericks who say they are about to clone babies has distorted the media debate and fuelled political efforts to ban all forms of cloning, both reproductive and therapeutic. In short, it has threatened to derail serious science.

One unsung and unlikely hero in the efforts to disentangle the complex issues so that my proposal, among many others from serious scientists, can be discussed in a clearheaded way is Bernard Siegel. This former owner of Florida Championship Wrestling and of the Miami Tropics, a minor league basketball team, entered the cloning story as a Florida attorney who had worked in family law. He turned his hand from everyday cases in Coral Gables to TV debates with Rael, the supreme leader of the Raelians (believers in aliens and supporters of human cloning). Siegel sued Clonaid, the organisation that claimed to have cloned a human being – 'Eve' – in December 2002, to have a guardian appointed for the alleged child. He reasoned that anyone reckless enough to create Eve was hardly going to make an ideal parent. At the time of writing, Clonaid had not come up with any evidence to convince scientists that Eve was cloned.

Siegel, dubbed 'the clone ranger', moved on to educating United Nations delegates about the ins and outs of cloning, reproductive science and stem cells. I was among the scientists who joined the advisory board of his non-profit organisation, originally called the Human Cloning Policy Institute (now the Genetics Policy Institute), to try to raise the level of the debate, and in particular to draw a clear distinction between reproductive cloning and the other uses, such as therapeutic cloning and the combining of cloning with genetic modification. We are still trying to make sure that life-saving and life-enhancing research is not ended along with attempts to clone babies.

At the heart of the uneasiness many feel about my proposal is the status of the human embryo: whether it counts as a full version of human life or lies somewhere between the status of a cell and that of a person. Even if one accepts this greyscale of life (and many do not), does the embryo still deserve respect? And if it deserves respect, would that prohibit its use in research? Do the ends – new treatments for horrific diseases – always justify the means? There are broader questions raised by my agenda. What do we mean when we condemn research as unnatural? What do we actually mean by a human life and a person? What is consciousness? Indeed, what is life? In the following pages, I hope to shed a little light on the forces that are now shaping our reproductive destiny.

Chapter 2

Cloning the Cloner

Three justifications for human cloning tend to crop up in public discussions more often than any others: the first is to overcome infertility, the second is to 'bring back' a dead child, and the third is to duplicate someone of astonishing ability or talent – to give the world another Mozart, Curie or Einstein. But how about cloning a person who is far from being an ideal male specimen (even to his wife and children), a bald chap with a beard who notched up a somewhat undistinguished academic record at school? Someone who is famous for his association with celebrity sheep? An individual who proves that scientists who do supposedly extraordinary things are far from being extraordinary, let alone a crazed Frankenstein or a Strangelove figure? An ordinary guy, like me. After all, if you prick me, do I not bleed? If you tickle me, do I not laugh? Of course I do.

In the years since my colleagues and I unveiled Dolly, I have been asked many times whether I would ever consider the ultimate act of vanity – cloning myself. There is, after all, a long and honourable history of medical scientists who have tested their ideas on the most curious of experimental subjects – themselves. Since the earliest days of medicine, researchers have swallowed microbes, injected vaccines, followed starvation diets and even performed surgery on themselves. Usually, researchers double as subjects to make their experiments both practical and ethical. But not in the case of cloning. Whenever I have been asked whether I would use myself as a guinea pig, I have had to explain how, even if the many technical obstacles could be overcome, and a mini-me, 'a little Ian Wilmut', could be born without risk, the answer would be the same. No. I have no intention of cloning myself. One Ian Wilmut is quite enough.

Living in my own shadow would be intolerable. Even though the reality is that a clone will be an individual living his own life, he would have to deal with the expectations of parents, family, friends, teachers and, of course, the media. That would be a heavy responsibility. Imagine what a broody teenager would make of being told that he is a genetic copy of a parent, let alone a cloner. Imagine what it would be like to know that you are the product of a scientific experiment. Imagine his irritation to learn from his pals that his father – also his brother – was called Wilf at school and perhaps that would make a good nickname for him too. But more than this, imagine living under the burden of feeling that, both medically and psychologically, the future would no longer be open, that his lifetime could follow the same trajectory as his father's. My little identical twin would be an individual, a person in his own right, but he would have to bear the heavy burden of growing up to feel like my duplicate. He would watch each serious illness that struck me with trepidation, wondering whether it was his destiny to suffer in the same way.

When a son or daughter lives in the shadow of a famous or high-achieving parent, there are often tensions. The unusual similarity and the unusual reasons for the birth of my clone, let alone my role in the birth and the controversy and furore that reproductive cloning has stirred, would inevitably create even greater and more unpredictable stresses. It seems inevitable that parents, relatives and friends would expect the clone to be an echo of the original person. (I have been asked again and again whether my children are biologists. For the record, Naomi is, but Helen is a mathematician, and Dean, whom we adopted, works in hotels.) This belief persists even though we know that each person is an individual and different from everyone else. There is still a strong expectation that we will be like our genetic forebears; this expectation would be greater, even overpoweringly so, for an unfortunate clone. Reproductive cloning runs counter to our present culture and would wreak havoc with family law. A survey of more than 1800 people for the UK Department of Trade and Industry, published in 2005, found that around two-thirds of respondents saw cloning as a 'bad thing'. Similar consultations have elicited an overwhelmingly negative response: people don't want it.

The Curse of the Clone

The psychological and social impacts of cloning have been explored by the psychiatrist Stephen Levick in his book *Clone Being*. Cloned children would have a confusing array of unusual relationships to relatives and suffer what Levick calls 'genealogical bewilderment'. The clone would be the identical twin of the donor, who could also be his parent. In a genetic sense, the parents of the person copied would also be the 'parents' of the clone. By contrast, the birth mother of the clone might make no genetic contribution if, for example, the child were a clone of a male or female partner.

In the absence of families with cloned children to study, Levick looked for relationships in modern life that might be comparable. He examined situations in which some of the effects of being a clone might be expected to appear, such as that of identical twins, who are natural clones of each other, and of children born through the use of *in vitro* fertilisation (IVF) and other reproductive technologies. The parents of these children, if they are candid about their high-tech method of conception, may constantly remind their sons and daughters that they are 'special'. Louise Brown, the world's first IVF child, used to wonder whether she would grow up to be normal: 'Am I going to grow up with two heads or something?' Schoolchildren would ask her how she fitted into a test tube. 'I knew that it wasn't a test tube, and I used to get fed up with explaining it then. They couldn't understand that I was actually born from Mum.'

Other important insights into the lot of the clone come from stepchildren. A stepchild is genetically related to only one parent, as would happen with cloning where there is one genetic 'parent'. Here the genetically programmed rule – help your closely related kin – may have an ugly mirror-image, according to Stephen Emlen of Cornell University. Stepchildren, on average, suffer higher rates of physical and sexual abuse and even death than children in intact families. Step-parents tend to invest less time and effort in the off-spring from their partner's previous marriage than they do with their own children. One study showed that the incidence of sexual abuse of stepdaughters was eight times that of biological daughters. Children in stepfamilies tend to leave home significantly earlier than do children in intact families. And stepfamilies are, on average, less stable than intact families. The incidence of subsequent divorce is higher in second marriages and increases with the number of stepchildren present. Levick expresses concern that similar risks might face a clone. The genetic parent of the clone may look upon the child as 'mine' rather than 'ours'.

Levick says that any parent motivated to clone himself or herself is bound to feel let down by the results. The narcissist is always disappointed in his or her children and their failure to live up to 'perfection'. He would surely feel entitled to expect a clone to be just like him, or his ideal. Many people have already had to struggle under the burden of unrealistic parental expectations. For a cloned child the pressure could be extreme, even though parents know from their own experience that nurture is as important as nature. 'A clone would be likely to face extra difficulties in negotiating every stage of psychological and psychosocial development,' Levick concludes. Despite his concerns about the likely psychological ill-effects of being a cloned child, he does not recommend prohibition: 'Cloning per se would not be wrong.' A ban, such as the one already in place in the UK, would stigmatise any cloned children. They would be made to feel a deep sense of shame even though they would carry none of the responsibility. And a ban would force cloning underground, making it even harder to help any clones.

In an attempt to undo their terrible loss, some parents might want to clone a dead daughter from one of the child's cells. Levick explored what might happen by drawing on the example of bereaved parents who try to deal with their loss by having another child in the hope that the new child will be a replacement for the one who died. A clone would regard herself as a replacement, a living memorial to the dead. She would have difficulty in developing her own unique identity, and would be unlikely to fill the shoes of the ghost. The potentially tragic results Levick foresees for such a child clone lead him to view any attempt to 'replace' the dead by cloning as unethical.

There are also good scientific reasons why clones will never be copies. Genes are not as powerful as many people think. A clone may have identical genes to those of a lost child, but that is not enough to ensure that he develops the same way. The influence of genes is modified significantly by that of the environment. Genes are in constant dialogue with their surroundings. They are in dialogue with the rest of the cell in which they reside, which is in dialogue with other cells in the body, which in turn is in dialogue with the world at large, through education and experience. This nested dialogue shapes the development of a fertilised egg into brains or brawn. Just as one can never really relive a moment, this dialogue can never be exactly reproduced. As I explain below, the personality of a clone will be individual. For this reason alone, cloning cannot resurrect a loved one.

Drop the Doppelganger

Many of the psychological overtones of cloning arise because many people still think – wrongly – that cloning is a way of copying another human being, just as a photocopier duplicates a document. They imagine that the process of cloning would seek to capture not just a person, but that person at a particular moment in time, perhaps even to go back to a lost moment, to try to defy what the novelist Milan Kundera called 'the unbearable lightness of being'. This cannot be done: even identical twins who have developed in the same womb and experience the same upbringing are never really identical. Genetic identity is not the same as personal identity, and selves, unlike cells, cannot be cloned.

The influence of environment, luck and circumstance would make it impossible for my clone to grow up in the same way as me. He would grow up in a different time and in a different place, eating different food and hearing a language decorated with different slang and idiom. He would have different best friends, different teachers, different lovers, and so on. Take two people, one of whom spent her formative years in the post-war austerity of the 1940s and the other in the relatively well-off and liberal years of the twenty-first century. They could not possibly think alike.

Parents of natural clones – identical twins who have come from the same human egg – often dress them differently and emphasise differences in character in a bid to make them true individuals. This may be helpful for the twins and for relatives and friends, but it is unnecessary. At the very start of life, the two embryos would have been implanted in different places in the uterus, or have drawn upon different parts of the same placenta. As a consequence, they would have had different access to food as well as different exposures to toxins and infections. They may have started off with the same genetic code, but that code is altered by errors and mistakes that occur as genes multiply in their bodies. The way genes are turned on and off by a process called imprinting can be patchy and different for each twin. Slight differences in the frequency of cell division or in the rate of cell migration may slightly alter the appearance of one twin relative to the other.

The brain would be affected in a similar way. Small changes in the way the nervous system wires up and matures might also influence ability or behaviour, although many other factors are also important in determining these characteristics. Identical twins are more similar than you and me are. But they are not really identical at all.

Plenty of evidence from cloning animals from cats to mules shows how environment shapes an individual as much as genes do. We cloned a quartet of young Dorset rams, Cedric, Cyril, Cecil and Tuppence, from embryonic cells that had been cultured in the laboratory. They were genetically identical to the same culture of Poll Dorset embryonic cells and yet were very different in size and temperament. Although they were all more aggressive than females, as one would expect, Tuppence was by far the quietest of the four. Small variations in cell division and migration result in differences in the pattern of black and white skin in cloned Holstein dairy cattle. The world's first cloned horse, named Prometa and unveiled in 2003 by Cesare Galli of the Laboratory of Reproductive Technologies, Cremona, Italy, could be distinguished from her sister/mother (unusually, she was the clone of the horse that carried her) by a white stripe. And so on.

As variation is to be expected in any species, pet owners should not expect clones of Tiddles to be identical to their late lamented cat. This was observed with the first feline clones, created at Texas

A&M by Mark Westhusin. Rainbow the cat was a typical calico with splotches of brown, tan and gold on white. Cc (for 'carbon copy'), her clone, had a stripy grey coat over white. And while Rainbow was chunky, Cc was sleek; Rainbow was reserved, Cc was playful; and so on. The prospects of making exact copies of Patch the cat and Spot the dog look bleak.

A Hard Act to Follow

Twins and animal clones show that because of tiny differences in the way genes are used in the body, the future of a clone is very much open. My own curriculum vitae illustrates the equally profound effects of happenstance and serendipity: the basic genetic clay of life can be pounded and shaped and moulded into endless variations.

My parents were modest, middle-class professionals. They both taught, and in those days could afford to buy their own home without the need for a special loan from the government. We enjoyed weekends and holidays in a caravan, touring Britain. My parents were especially fond of exploring the remoter reaches of Scotland, where little-used single-track roads had grass growing down the middle.

Although I was born at Hampton Lucy, near Warwick, I was brought up in and around Coventry, where my father taught. Before World War II, Coventry had been one of the best-preserved medieval cities in Europe. But when I ran around its streets as a boy, it bore many scars. It had been the industrial heart of Britain's wartime manufacturing, making it a major target of Nazi bombing raids. Unlike the village where I now live, it offered no panoramas, vistas or horizons. I felt claustrophobic, and would come to crave wide-open spaces.

When I was six we moved to Shipley, in the West Riding of Yorkshire, and I attended the school where my father taught, in

Saltaire, the model industrial village built by the Victorian philan-
thropist Sir Titus Salt. While visiting my grandfather in the coastal
resort of Morecambe at the age of ten, I met an inspirational figure,
a merchant navy captain. My future was now quite obvious: I
wanted to leave Shipley and sail the seas. Over the next few years,
my interest in the ocean grew into an obsession. But there was one
problem – a mutation of a gene on my X chromosome.

I am among the 7 per cent of the male population who cannot
distinguish red from green. I lacked the right stuff for the navy.
When it comes to teenagers, that little word 'no' is a poor persuader.
Rather than criticise my chosen career path, my family adopted a
more subtle way to convince the fourteen-year-old me that life on
the high seas was unsuitable. An uncle who happened to be an opti-
cian sent me to the docks so that I could learn first-hand what
would happen if I could not distinguish a red navigation light, for
port, from a green starboard one. I quickly realised that Captain
Wilmut would be a danger to himself, to his crew and to other
ships. Today, my clone would find that the navy is happy to accept
the colour-blind, but only in certain posts. My clone could end up
sailing the seven seas rather than being a prisoner of the international conference circuit who flies across the world's waters and
never sails upon them.

If my clone knew much about my upbringing, he would be
apprehensive about going to school. By the time I faced my first
qualification exams, at the age of sixteen, I had moved to
Scarborough and, at a critical time for my education, had been off
school with appendicitis. I managed to get only four O levels and
had to retake them. When the next academic milestone loomed –
A levels – I was told that I was falling behind in chemistry. I made
up for lost time and passed chemistry and zoology. But I failed
botany.

I would love to be able to tell here some homespun story about
how, despite this hammer-blow to my ego, I picked myself up, stuck

out my jaw defiantly and told my school friends that one day I would do something that would give the president of the United States pause for thought. But I had no delusions of grandeur and saw myself as an average sort of schoolboy. My botanical setback merely underlined this fact. Perhaps, however, this low-key start to my career would take the pressure off my clone to perform.

Would my clone be lucky, like me, and fall into the arms of the love of his life at school? If not, he would lack what I consider to have been a key ingredient of my life, my emotional motor that has helped keep me going. I first met Vivienne when we were studying for our A levels in Scarborough. She passed her A levels in mathematics, physics and chemistry. She still motivates me today. I feel I know her better than I know anyone else. Yet it is hard for me to imagine what would go through her mind if she had borne my clone and found herself looking after a schoolboy who looked just like the one she had fallen in love with decades earlier. How would she regard him? Some observers, including Levick, have even speculated that cloning would raise the spectre of incest.

Four decades ago, after my undistinguished performance at school, I found myself working on farms and then attending agricultural college at Nottingham University. Farming was my next love, not least because I wanted to breathe fresh air. But I discovered that you need to have a good head for business too, and in this important respect I was lacking. At Nottingham I became gripped by a subject that has fascinated people since the days of the ancient Greeks – embryology, the study of how a minute speck turns into a living, breathing animal – and studied with G. Eric Lamming, an authority in the field of animal reproduction with a down-to-earth nature. A farmer at heart, Lamming established Sutton Bonington as one of the top animal science research institutions in the world, and that is where Keith Campbell would end up as a professor.

I enjoyed working in the laboratory at Nottingham, and in my second year I decided to write to nearby research centres to find myself a summer job. A ripple of increasingly plaintive pleas for work spread out from my digs in Sutton Bonington to labs at progressively greater distances from Cheltenham, the Regency spa town in Gloucestershire where Vivienne was working and where I wanted to be. This led to a lucky break that would change my life. I won a scholarship to work at the Animal Research Station, Cambridge, which made major contributions to the development of methods for freezing semen, producing pregnancies by transferring semen into cows (artificial insemination), recovering bovine embryos, growing them in the laboratory and transferring them between different mothers. That summer, Chris Polge, one of the great pioneers of animal reproductive science, both as an entrepreneur and as a researcher, took me on for a month or two. That marked the real start of my career.

Frostie

Chris is a far-sighted scientist and a patient teacher, and still a generous friend. During that first summer I worked for him, the station was host to Professor B. N. 'Billy' Day from the University of Missouri-Columbia. He was visiting Britain to work on the oestrus cycle of the pig, and how to control it, and a fine steak and rib barbecue was held in his honour by Chris and his wife, Olive, at their home in Girton. It was around this time that I had my first experience of wealth (at least what I thought of as wealth) and advanced office technology (for the mid-1960s): I caught my first glimpse of a £20 note, which I had been given to pay for the use of a photocopying machine, creating in me the quite misleading impression that scientific research was a way to make money (certainly not working for a British government lab, alas). I spent hours copying experimental records so that Chris and Billy could each have copies of the records once Billy returned to the United States.

While I had to do my fair share of routine laboratory work, including looking after the pigs and cleaning up in the lab, I had my first chance to work with sperm and embryos. I studied living embryos for the first time, watching them as they began to organise themselves into a bag of skin with its associated bones, muscle and plumbing – the foetus – and its ancillary equipment: the placenta, umbilical cord, and so on. I also learned to recognise the signs that would reveal which eggs were fertilised and developing normally.

Like many other students, I learned as much in the coffee breaks as I did in the laboratory. Sometimes the chat was about cricket, a subject close to the heart of Chris Polge. At other times it was about gossip and personalities, usually of the scientific variety. But most of the time we discussed reproduction. Just exactly how does a mated sow know whether she is pregnant or not? It turns out that a pig maintains a pregnancy only if she has four or more foetuses developing inside her. If there are fewer, she comes back into heat to have a chance of producing a larger litter. I was hooked by such questions. I knew then that, if I could, I wanted to build a future for myself in research.

After the summer work I found myself back at university, looking for a research post. I wrote to Chris later that year asking whether he would give me a reference. I did not ask for a job because I knew there were none to be had. Here again, any effort to create a 'mini-me' would find it hard to replicate what happened next. Few people have a career trajectory that goes in a straight line. But at this crucial point, mine did. Chris surprised me by asking me back, offering me a postgraduate position funded with a scholarship from the Pig Industry Development Authority. It turned out that I had been lucky because of another's awful misfortune: one of Chris's team had committed suicide. My feelings were, of course, mixed. I was uneasy about the terrible circumstances, but this turned out to be an amazing break without which my career might well have fizzled out there and then.

After finishing my studies and achieving a good second-class honours degree, I returned to the Animal Research Station where I found myself studying for my doctoral thesis at Darwin College, Cambridge, working on the freezing of boar semen. Most important, I was again under the spell of Chris, who, among his many achievements, had done the pioneering work in the fifties that showed how a small molecule – glycerol – can protect frozen cells. He also had a management style that suited me: he was always available to challenge my ideas and offer advice, but he left me to get on with my research.

That first project would set a pattern for my subsequent work. Finding out how to successfully freeze semen requires a systematic approach that would later come in handy for perfecting cloning. To freeze sperm in such a way that they can later be revived depends on which protective chemicals are used, the rate at which the sperm are cooled, and many other factors – so many, in fact, that it is easy to get lost in a forest of permutations.

I had the good fortune to be taught by a Hungarian visitor to Chris's lab, Steve Salamon, who gave me many insights into the design of effective experiments that could provide a clear path through this jungle of variables without descending into the chaos of pure trial and error. Inspired by Salamon and with equipment designed by John Farrant at the Medical Research Council in Harrow, I tested several different freezing regimes using what I called my 'four-seater loo'. Each 'loo' was a flask of liquid nitrogen in which embryos were cooled at rates varying from a fraction of a degree Celsius per minute to 100 degrees per minute. It was probably around this time that I first became aware of cloning, notably the efforts to multiply rabbits by Derek Bromhall in Oxford, who would become one of the first victims of human cloning (more of him in the next chapter).

After I finished my Cambridge University doctorate, in 1971, I was lucky to be able to carry on working with Chris Polge, this time funded by a fellowship from the Milk Marketing Board. Now defunct, this organisation supported all aspects of milk production and distribution, including the breeding of superior cows. Research on cattle by biologists such as Chris and geneticists based in Edinburgh had given Britain some of the finest (in terms of being productive) dairy cattle in the world. I was able to work with Tim Rowson, the vet at the station, from whom I learned the techniques of surgery, a skill that would one day come in useful for cloning.

We were the first to succeed in freezing a calf embryo, thawing it, and transferring it to a surrogate mother. The challenge was to stop ice crystals from forming, which would fatally damage the embryo. During cooling, as ice forms in the solution around the cells of an embryo, the salts in the remaining solution become more and more concentrated. As a result, water will flow out of the embryo into this solution, by a process called osmosis. I established the cooling rate that stopped ice crystals from forming in the embryo. Thawing must also be controlled carefully. Because it takes much longer to obtain the embryos and test their viability, this was a laborious project, but it was worth the effort.

The result was the world's first 'frozen calf', a red and white Hereford–Friesian cross. We decided to call him Frostie. The events that led to his birth in 1973 were filmed by a Dutch TV crew, who camped out in a barn for several days to be sure to witness his arrival. When Frostie was delivered by Tim Rowson, we celebrated with champagne, a suitably extravagant display prompted by the presence of cameras. As the *Daily Mail* put it, 'Ice-Age Calf Weighs In'. I also got my first experience of dealing with the media. Fortunately, while I was driving to my first ever live TV interview in Norwich, I heard a science correspondent discuss Frostie on BBC radio and give some useful sound bites on his significance for agriculture. Even with these tips on how to be succinct and comprehensible, I was still terrified as I spoke from a studio in Norwich to the presenter of the *Nine O'Clock News*, which had a vast audience in those days.

Thanks to my time with Chris Polge in the 1960s and '70s, I had a background understanding of reproductive physiology, the basic techniques needed for this area of research, including surgery, and the management of experiments that were more likely to end in failure than in success. I began to think about life and death: when we attempted to revive the embryos from their frozen state, some were dead and unresponsive. But what was their status before they were thawed? And, of course, I began to appreciate the pressures of dealing with the media. This experience would prove useful when it came to Dolly.

Roslin, Pharming and Beyond

Frostie gave me a reputation, one that presented new opportunities. In 1974 I joined the government-backed Animal Breeding Research Organisation, ABRO, in Scotland. Initially I was based at laboratories on a farm at Dryden, near Roslin, alongside the sheep sheds and pig barns, but in 1987, as part of a major reorganisation, we were all moved closer to the village of Roslin (*ross*, 'point', and *lynn*, 'waterfall'), where a ruined castle lies, a mining industry once waxed and waned, and the surrounding lowland hills were grazed – and still are – by flock upon flock of Scottish Blackface sheep. The village is also the location of the fifteenth-century Rosslyn Chapel. Perched above the North Esk with its richly carved interior, it may, according to some believers, contain the Holy Grail, or the Ark of the Covenant, or part of the actual cross on which Christ was crucified.

What became the Roslin Institute in 1993 merged two research outfits, one concerned with poultry and the other with sheep, cows and other farm mammals. Today the Roslin is backed by the government's Biotechnology and Biological Sciences Research Council and inhabits a long, glassy building with steep roofs, mostly one storey – anything taller might be unwise, since the ground beneath is riddled with disused mining tunnels. The Roslin remains at the forefront of animal genetic research.

When I started at the Roslin in 1973, I focused still further on the basic repertoire of reproduction – eggs, sperm and embryos – with the aim of understanding the causes of infertility. At one point the institute was threatened with having its spending cut by two-thirds. I had carved out a reputation as a research embryologist, so when government funding went into reverse and the ABRO con-centrated on genetics, I was able to keep my job, despite being a physiologist. The political dogma of the day was that scientific research must be utilitarian so that it could pay its own way – a philistine attitude which shows little understanding of the complex relationship between fundamental 'blue skies' research (of the kind that usually seems pointless to a bean-counter) and the development of innovative products. Under the then prime minister, Margaret Thatcher, research was being squeezed: on one side, we were encouraged to do work which could have a pay-off in the foresee-able future; on the other, any research that looked close to profitability should be cut adrift of government funding and be paid for by industry.

At this time, molecular biology was becoming fashionable. The new director of the ABRO, Roger Land, brought my work on embryo death to a close in 1982 and told me to work on the genetic alteration of animals, which in those days was a tedious slog that usually ended in failure, but which did draw upon the skills of an embryologist. I had to end my research abruptly: either that, or resign. I was unhappy with the thought of merely recovering embryos from ewes and adding DNA, and thought long and hard about leaving. There were then many redundancies, and the ABRO was under threat. Agricultural science in Britain felt almost doomed, and I discussed with Vivienne what I should do next.

The countryside and way of life intervened. Vivienne and I loved the Scottish borders. We played an active part in the life of the local village, where I was a member of the farmer's club (of which I eventually became president). The surroundings were spectacular, and we walked our dogs in the hills. By now we had evolved into Lowland Scots, and decided to stay on, not least because we thought that the children would benefit too.

I agreed to become a genetic engineer. I also told Roger that, once I had mastered genetic transformation, I would look for more efficient ways to achieve it. It would remain a struggle for some time: between 1989 and 1996 my colleagues at PPL Therapeutics used 2877 ewes to create 56 viable lambs with new genes. I would come to realise that Roger wanted to be involved in everyone's decisions, rather than giving them the room to get on with their jobs. I never saw eye to eye with him in the same way that I had with the relaxed and affable Chris Polge. Nonetheless, by the mid-1980s I had become involved in the ABRO's effort to genetically alter a range of animals including mice, sheep and pigs.

Genetic modification is, in one sense, as old as domestication. All varieties of domestic dogs, from Dalmatians to chow-chows and Rottweilers, have been created through selective breeding – that is, breeding from selected individuals. The same is true of cows and sheep. Now we wanted to find a way to alter the genetic make-up of animals with unprecedented precision. And we wanted to do more than improve the performance of farmyard animals. The idea was to give these animals an entirely novel role and convert them into living drug factories to make the human proteins factor VIII and factor IX used to treat haemophilia, and the enzyme AAT (alpha-1 antitrypsin) used to treat lung disorders such as cystic fibrosis. At the time this was a new field, a combination of pharmaceuticals and farming, aptly dubbed 'pharming'.

Old-Fashioned Pharming

During these pioneering days, pharming was tiresome and repetitive. Unfortunately, high science and lofty aspirations gave way to factory work and repetition. The idea was simple; the practice was monotonous. Each gene in the body is, in effect, the recipe for a protein (it is, of course, more complex than this, since genes can be shuffled in various ways to make more than one protein). All one had to do was use a tiny hypodermic needle to introduce a gene into the two metres or so of DNA that was packed into every cell. The way we did it – using a method first demonstrated on mice by Francis Ruddle and Jon Gordon of Yale University in 1980 – was as crude and obvious as that. We squirted in naked DNA and hoped that it would be used by a developing embryo. By the time I started this work, the public had been alerted to its possibilities by one of the great pioneers of reproductive science, New Jersey-born Ralph Brinster, who had funded his education with a poultry business at the age of thirteen.

With his postdoctoral fellow, Bob Hammer, Brinster showed how one could make mice grow bigger – into 'supermice' – by putting in a growth hormone gene. Then, with Ursula Storb, Brinster proceeded to 'turn on' genes implanted in specific tissues, a handy trick if you want to make drugs in milk, for example, which is more elegant and efficient than doing it in the lab. A recluse with a moustache who shuns most meetings and committees, and a Korean War veteran, Brinster has since done important work on sperm stem cells, on implanting them to restore fertility and on their genetic modification. Hammer has remarked that Brinster's approach to complex scientific problems was always the same: 'It's almost like he's waging war. He sees the problem, breaks it down, then goes after it. He's probably the most tenacious individual I've ever met.'

After introducing genes into sheep embryos and introducing the resulting genetically modified (GM) embryos into sheep, a thousand in all, I had first-hand experience of how this new method of genetic modification was hopelessly and maddeningly inefficient. To do a thousand embryo transfers successfully requires a military-scale operation with precision timing, day in and day out. The methods of genetic modification we used in those days were not only inefficient but also limited: one could add a gene but not fix an existing faulty gene. Imagine the frustration of a car mechanic who can only add new components to a faulty engine, never repair or replace existing parts.

As we started this work, it became clear that one of my own limitations would always prevent me from carrying out the micromanipulation involved in genetic modification or cloning. The dark art of embryo manipulation was beyond me. I had then, and still have now, a tremor that makes me hopeless at executing the tiny, delicate hand movements required to operate a microscope, bending fine pipettes so that they can reach into the chamber in which the embryos are held, and injecting genes – that is, if you could see the nucleus of the early embryo into which they have to be inserted.

Embryos are in fact opaque, but thanks to painstaking observations by my colleague Paul Simons we found that in seven out of every ten embryos we could locate the nucleus by patiently rolling the embryo around and searching for a little clearing, where the DNA resides. Furthermore, there is not one nucleus but two. This feature of the earliest stage of life might well surprise anyone who knows only the school textbook account of conception. Look closely at a newly fertilised egg under the microscope and you will see that within the early embryo the egg's DNA remains separate from the DNA imported by the sperm.

Inside most cells, the genetic instructions are often to be found in a package called the nucleus. But here, at the outset of a life, the DNA actually sits in two 'pronuclei' (think of a chicken's egg with two yolks). The existence of two nuclei, where female and male DNA sit in splendid isolation, raises a question about what we mean by an 'individual' when we have an entity with a split genetic identity. In the pioneering days of *in vitro* fertilisation, it led to heated debate in the scientific literature about when the fertilisation of test tube babies actually happens. This debate has still not been resolved.

The good news for researchers in reproductive science is that even at this earliest stage of life there is lots of chemical machinery to cope with insults and errors. I suspect that the machinery includes some proteins that organise the pronuclei. In cloning, we now rely on these proteins to reorganise the nucleus of a donor cell. This same machinery can also cope with foreign DNA, allowing genetic manipulation to take place. You might think that the foreign DNA should be put into both pronuclei. In fact, genetic engineers inject the DNA into just one of them – there is a limit to the insults that nature will tolerate, and to put new genetic instructions into both would be too disruptive and kill the embryo.

The results of this simple genetic manipulation are very much hit and miss. Of all the embryos injected with foreign DNA, only around 1 per cent become live animals carrying a functional added gene. In fact, the reason for this low efficiency is more subtle than simple hit or miss, as I shall explain in a later chapter. Nonetheless, we made steady progress. We produced our first genetically altered (transgenic) lamb in 1985, just as the Brinster lab was unveiling transgenic rabbits, sheep and pigs. A company was set up to produce, in the milk of sheep, human proteins needed to treat disease. From this work came our first breakthrough: Tracy, a Scottish Blackface sheep who was born in 1990 and was probably the most famous sheep in the world until the birth of Dolly. As if to underline her commercial value, she was owned by our neighbouring biotech company, PPL. Tracy's milk was rich in the human protein AAT, which was used to treat cystic fibrosis and emphysema. In every litre she made 35 grams of AAT. To put this extraordinary feat into context, four of our other transgenic sheep managed to make only a gram or so. Tracy was a living drug factory.

The Drive for Dolly

The creation of Tracy was, despite all the effort, a relatively simple feat because it rested on adding a single gene. Most traits of interest to breeders depend on constellations of genes. Let us say that you have a one-in-four chance of getting a working gene into a safe part of the genome (the complete sequence of DNA carried by an organism), which is actually a far higher success rate than is currently possible. Imagine that you understand the biochemistry of a trait and that you want to enhance it by tinkering with ten genes. To succeed, back then you would have needed about a million GM embryos. Of that million, a quarter of a million would have one of the desired genes, a quarter of those 250,000 would have two, and so on. Just one in a million would have all ten. Traditional genetic modification was, to put it mildly, inefficient when it came to altering complex characteristics.

And that is why I helped to create Dolly. Many people think that I was motivated primarily by cloning for its own sake, to take a superior animal and make identical copies of it. I did not seek simply to produce facsimiles of existing creatures, not least because this is impossible, as highlighted above by the impact of serendipity, randomness and chance on my upbringing. I am not concerned primarily with making multiples of elite livestock (though that was a motivation) and still less do I want to clone human beings. That has never been on my agenda. It was what other people, especially in the media, read into my work. For me, cloning has always been a tool of science – finding out how cells work – and a means of improving the technology of genetic transformation of animals.

The original inspiration for the research that created Dolly dates

back to work conducted early in the 1980s by Martin Evans and
Matt Kaufman at Cambridge. They showed that it was possible to
isolate lines of embryonic stem cells from mice by means of so-
called feeder cells, which provide the stem cells with support (in the
form of a cocktail of as yet unknown cell growth factors and other
trace chemicals). When an embryo is a few days old – what embryologists call a blastocyst – and ready to implant, it is hollow and
contains a clutch of cells called the inner cell mass. Evans and
Kaufman found a way to grow and multiply cells from the inner cell
mass of a blastocyst – the part that will turn into the foetus and
most of the placenta.

These embryonic stem cells are rather strange. You can grow
them indefinitely in the laboratory, and that provides lots of opportunities to genetically alter them. They also retain the unique ability
to form any of the many different tissues that make up an adult
animal. Mix these stem cells with the cells of an early eight-cell
embryo, or inject them into a blastocyst, and you find that the
resulting offspring is a blend of the descendant cells of both the
embryo and the newly introduced stem cells. Scientists call this
blended creature a chimera (in Greek mythology, the chimera was
a monster that had the head of a lion, the body of a goat and the tail
of a dragon). Despite this extraordinary potential, you cannot grow
a stem cell into a whole mouse, as you can with the very earliest
embryonic cells. This is partly because stem cells are too small to
form an embryo, which is much bigger than other body cells. But
you can grow them into particular cell types, such as nerve, skin or
bone cells for treatments, as we shall see in a later chapter.

By blending stem cells with an embryo to make a chimera, one
can turn the stem cells into the cells that make eggs and sperm: the
mouse will then produce eggs and sperm that are the genetic offspring of the cultured stem cells. If you had genetically altered the
embryonic stem cells, the chimeric mice would also produce eggs
and sperm with that alteration. This approach has been the basis of

an extraordinary range of experiments on mice to study the role and function of newly identified genes. Because mice take relatively little time to become sexually mature – four to five weeks – the production of chimeras did not present a problem when it came to creating GM rodents. But it would have taken us decades to achieve change in relatively long-lived creatures such as cattle or sheep.

The idea of making Dolly was born when it occurred to me that genetic modification would be so much easier and more efficient if we could first take stem cells from an early sheep embryo, multiply them to form millions of cells, then add DNA to those cells and select only the ones in which the process had succeeded for the use in cloning – nuclear transfer – to make animals. Whereas it is trivial to add genes to millions of cells in a test tube, then find the cells in which genetic modification has worked properly, it would be impossible to achieve the same feat for classical genetic modification of millions of embryos. And it meant that, unlike the earlier scattergun approach, this method was much less likely to produce sick GM animals: the industry was not going to benefit if the animals were hurt in any way by genetic modifications.

Only once you have identified the embryonic stem cells in which your genetic transformation has been successful can you go on to convert those GM cells into GM embryos. Such multiplication of animals from cells is what cloning is. At the time I thought of doing this, however, cloning was routine only with bacteria and plant cells. I wanted to do the same with the cells of an animal. And to do this, I would have to extend an effort to clone a menagerie of cold-blooded creatures by a number of scientists that stretched back decades – an effort that had already stimulated an unprecedented amount of debate, controversy and anxiety.

Chapter 3

A Brief History of Cloning

Dolly the lamb marked the culmination of more than a decade of intensive research at the Roslin Institute by me and my colleagues. But the credit for this reproductive feat cannot be given to us alone. The achievement of her birth rested on a sturdy foundation of skills, understanding and invention that had gradually been built since in the nineteenth century by scientists around the world. And it relied on a zoo of species. The early clones that blazed the trail to Dolly were not just sheep but also rabbits, mice and cattle. Perhaps the most important pioneers of all were cold-blooded clones of sea urchins, salamanders, frogs and toads.

Despite this long history, the path to Dolly was far from straightforward. There were diversions caused by mistakes, muddles and misunderstandings. There were rivalries and controversies, even accusations of deceit. All the while, many speculated on the implications of advances in reproductive science. These ideas were embellished in novels, movies and other imaginative creations. Fictional fascination with cloning has rarely focused on scientific fact, but usually instead on issues of identity and how the sanctity of life will be challenged when 'ditto machines' of one kind or another create 'cookie-cutter humans'. This obsession has led to endless confusion about what is possible and what is not.

While some people fear (wrongly) that a clone would be a duplicate not just of a person's DNA but also of their consciousness, others fear (again, wrongly) that clones would be soulless replicas. Popular culture has become obsessed with the idea that, in a world in which individuals are easily duplicated by cloning, life will be cheap and a person will be reduced to a replaceable unit, a mere component in the machine of society.

Yet in one respect, science fiction is already science fact. Cloned humans have been around since the dawn of humanity. A few days after an egg has been fertilised by a sperm and reaches the blastocyst stage, it has to 'hatch' from its rubbery shell, the zona pellucida, so that it can implant itself in the wall of the uterus. During this process, one individual can become two – the embryo can split. Sometimes the process of twinning does not occur properly, leading to conjoined twins. One striking and ancient depiction is a marble statue of two women linked at the hip, excavated from a neolithic shrine in Anatolia. Down the ages, great minds have pondered what conjoined twins tell us about development. The Greek polymath Aristotle (384–322 BC) mused upon whether they arose from two embryos that had combined or from one that had split: fusion or fission?

Intriguingly, we now know that both processes can occur in the womb. On rare occasions two eggs ovulate, and both can become fertilised; and sometimes those embryos can fuse to create a chimera. Alternatively, an embryo can undergo fission and split into two. There are variants on the theme of identical twins, depending on the degree of fission and how much plumbing they share in the womb. Some twins still draw on the same placenta. Others will develop with separate placentas in different parts of the womb; of these, some will have their own membranes and others will share one.

Embryo cleavage happens in the rest of the animal kingdom too. The nine-banded armadillo produces four genetically identical progeny from each fertilised egg. This 'natural' process of fission was also exploited in scientists' very first attempts to create a clone, before the other type of cloning – by the process of nuclear transfer – was developed to do the job, the first creation of a normal fertile adult animal from a body cell of another adult being the most dramatic example of all.

Pioneers of Cloning

Early cloning used creatures that, in a way, have been much more important and influential in this field than sheep: sea urchins, salamanders, frogs, and other creatures that have eggs so big that they are easy to study. They offer another advantage: whereas the Dolly effort relied on a labour-intensive combination of hormones and surgery to harvest a few eggs from a ewe for cloning, frogs make fabulous quantities of eggs – thousands and thousands – and deliver them in the convenient form of spawn. Frogs and their amphibian cousins are gifts for the curious cloner.

One of the first to try embryo splitting was the German Hans Driesch (1867–1941), the son of a gold merchant, who became a philosopher, biologist and master experimentalist. He took a two-cell sea urchin embryo and shook it apart. Both cells developed into complete individuals. Then he dismantled a four-cell embryo, and quads were the result. This was probably the first successful human-assisted animal cloning from a single embryo. In some ways he set the pattern for future work. Driesch, like me, was driven by an urge to explore basic science, not to clone for its own sake. And like me, he was in awe of the process of development in which a fertilised egg divides again and again, differentiates and becomes more complex, as cell layers slide over each other and fold in upon themselves in highly choreographed auto-origami to create an urchin, a frog or a man. Unlike me, however, Driesch was convinced that the ultimate explanation for development lay outside the boundaries of science: he believed that a 'vital force' propelled the miraculous transformation from fertilised egg to adult.

Driesch was the last great spokesman for this idea of vitalism. Although it still seems miraculous, we now understand much more about the origins of a life. Like other developmental biologists, I see no evidence of an invisible hand at work. The laws of physics and chemistry are sufficient to explain the behaviour of the molecules in a living cell. There is no room or need for a vital force.

The pioneers of cloning were trying to understand the basic details of the biology of development. These first ventures into the field were concerned with whether humans were fully formed at the moment of fertilisation, and just grew larger, or whether they developed from a single cell into the complex arrangement of billions of cells in a person. By the time Driesch came on the scene, there were two conflicting theories about how cells gradually differentiated from the blank slate of an embryonic cell to have a dedicated purpose, whether as a brain cell or a liver cell: the cells either lost all genes except those relevant to their specific task, or they kept the entire genetic recipe – the genome – and selectively turned genes on or off.

Driesch's experiments were regarded as refutations of a then prevalent idea first suggested by another great German, August Weismann (1834–1914), who worked at the University of Freiburg. In the 1890s, Weismann challenged Jean-Baptiste Lamarck's theory that traits acquired by parents during their lives could be passed on to their offspring (a theory overtaken by those of Charles Darwin and Gregor Mendel) by cutting off the tails of twenty-two generations of mice (1592 in all), demonstrating that the injury was not passed on. He became an influential figure in the history of eugenics.

Weismann believed that the genetic information contained in a cell would peter out as the cell underwent differentiation. If he was right, then separating and growing each cell from a two-cell embryo could create only half a creature. But the embryo-splitting experiments that Driesch conducted in 1892 showed how, a few divisions after fertilisation, embryonic cells retain the ability to turn into any type, from heart cells to egg or sperm cells, or even a whole individual (biologists say that they remain 'totipotent').

In the light of subsequent work, we now know that single cells taken from sheep embryos at the stage when they have two cells are each equally able to become lambs. The same goes for other livestock. One might expect the calves or lambs that result from fission to be smaller than usual, but recent work has shown they are actually of normal size. Part-embryos put on a growth spurt to get back on the normal developmental track. Conversely, when embryos are lumped together, the rate of growth declines accordingly.

This form of cloning, which has a long history, suffers from the law of diminishing returns: there is a limit to the clones that can be divided out of a single embryo, and this limit varies from species to species. The leading British embryologist Anne McLaren, while working in London with colleagues, found it a struggle to obtain pups from mouse embryos that had been divided in half at the two-cell stage.

This method has been used to duplicate our close relatives. A rhesus monkey was the first to be cloned this way, by Gerald Schatten while he was at the Oregon Regional Primate Research Center. He took rhesus monkey embryos at the eight-cell stage and separated them into four two-cell groups, creating a group of identical quadruplets. Two embryos were put into each of two surrogate mothers. Both became pregnant, each with only one of the two embryos. One mother miscarried, and the other gave birth to Tetra.

The same method could be used on people too. A much

misrepresented experiment using early human embryos in 1993 by Jerry Hall and Robert Stillman of George Washington University suggested that fission could be used to boost the numbers of embryos, and thus the chance of success, of *in vitro* fertilisation. Although this type of cloning is 'natural', and although their aim was reasonable, the research caused a furore. This application is unlikely to catch on because of public fears stirred by the 'c' word.

Nuclear Clones

The first glimpse of the nuclear transfer process that we used to create Dolly came decades after the first attempts at twinning in the laboratory. One element of nuclear transfer arose from research by the German-American Jacques Loeb (1859–1924). While working at the University of Chicago, Loeb showed how to simulate fertilisation in sea urchin eggs, a process we now call activation. Another element came from the pioneering and innovative German embryologist Hans Spemann (1869–1941), the son of a publisher who became, among other things, director of the Kaiser Wilhelm Institute of Biology in Berlin. In 1919 he was appointed professor of zoology at the University of Freiburg-im-Breisgau, a post he held until he retired in 1935. During his influential career, Spemann helped to establish what he termed *Entwicklungsmechanik* (what we would call developmental biology) to investigate how a fertilised egg develops into an adult organism. He was awarded the Nobel Prize for Physiology or Medicine in 1935, and remained the only embryologist to be so honoured for half a century.

With Hilda Mangold (who died at the age of twenty-six, after her dress caught fire while she was refilling a paraffin stove), Spemann discovered remarkable details of development by transplanting tissue between newt embryos to create conjoined-twin newts. He also conducted the first nuclear transfer experiments on the newt, another favourite subject for this kind of work because of its large and easy-to-manipulate eggs. For aficionados, Spemann is the true father of cloning. In the winter of 1896/97, while convalescing in a sanatorium after being infected with tuberculosis, Spemann read Weismann's theory of heredity and development and was inspired to investigate for himself. In 1914 he published an account of a fascinating experiment. The enterprise culminated in 1938 with his masterwork, *Embryonic Development and Induction*, in which he described a rudimentary form of nuclear transfer.

Key to his stunningly effective technique was something very simple: a delicate hair from a blonde infant less than nine months old. (After his death, a lock from Margrette, his baby daughter, was found in an envelope tucked into his 1899 file holder.) Working on the slippery eggs of salamanders – hundreds in all – while hunched over a binocular microscope, he used the hair as a noose to constrict a newly fertilised salamander egg. In this way he was able to divide the cytoplasm – the part inside the cell membrane and outside the nuclear membrane – to create two sections of the embryo. In one section of the resulting dumbbell-shaped embryo there was a nucleus, containing DNA, while in the other there was only cytoplasm, other cellular material. After the nucleated side had divided four times, creating a sixteen-cell embryo, Spemann loosened the hair and allowed one of the sixteen nuclei to slip back into the separated cytoplasm.

Spemann had created a new cell made up of some of the original egg fluid and a new, more mature nucleus. This new cell began to divide as well, and by tightening the hair loop again Spemann broke apart the two embryos. It was an extremely elegant experiment. Spemann had moved the original nucleus out of a portion of an egg and then put a more developed nucleus back in its place. He had shown that the nucleus retained the ability to turn into any type (i.e. it was totipotent) after undergoing four divisions. He had invented a new way of cloning. What a neat trick it was – all the more so, given that it was accomplished with two tweezers and a baby's hair. This experiment had an astonishing outcome: twin salamanders, one moments younger than the other. Spemann called the process that gave rise to them 'twinning', but we know it today by a different, more loaded word: 'cloning'. Spemann had produced the first documented *in vitro* animal clone by nuclear manipulation.

He wanted to go much further. In his 1938 book, Spemann proposed a 'fantastical experiment', a *gedanken* (thought) experiment that would lay the basis for Dolly. It would be salutary, he said, to take the nucleus of a differentiated cell – even from an adult cell – and place it in the cytoplasm of an egg whose own nucleus had been removed (enucleated, as we say today). But he did not carry out this experiment, for practical reasons: although he could work out how to isolate a nucleus, he did not know how to introduce it into an egg that had been drained of its own DNA, how to perform what we now think of as nuclear transfer. In the context of the science of his time, this proposal really was 'fantastical'. It was as outrageous as asking now whether we could reconstruct a person after ripping the brain from one body and thrusting it into the emptied skull of another. Indeed, Spemann himself was so in awe of development that he suspected it depended on more than mere physics and chemistry.

Even though we know and can do so much more today, Spemann was truly remarkable and exceptional. He was working by hand on small eggs, whereas we use instruments that can execute precisely controlled small movements. Not only was he an experimental genius, he was thinking differently. After all, he had the basic idea for the Dolly experiment. In the next decade, Spemann's challenge caught the eye of Robert Briggs, who had worked in a shoe factory and also earned money as a banjo player in a dance band before a high school teacher sparked his interest in biology. He later recalled, 'At the time it never occurred to me that I would become a biologist. I didn't even know that one could earn one's living this way.'

By 1949 Briggs was a researcher at Lankenau Hospital Research Institute in Philadelphia (later the Institute for Cancer Research, and now the Fox Chase Cancer Center). He was studying the role of the cell nucleus and its cargo of chromosomes in development. What governs how the genes on a chromosome are used by an embryonic cell to lay down its identity? A colleague, Jack Schultz, suggested that he might attempt something along the lines of Spemann's fantastical experiment, because nuclear transfer could probe whether genes are turned off during development or whether they are lost for ever.

When Briggs applied to the National Cancer Institute to carry out nuclear transplantation, his proposal was deemed a 'hare-brained scheme' with little chance of success. He reapplied, and this time, after a site visit by an institute representative, the application was approved with a large enough grant for him to hire a dextrous assistant – Thomas King of New York University, an expert on micromanipulation who was working all hours to support his family.

For their study, Briggs and King chose the common American leopard frog and its big, one-millimetre eggs. Using microscopic pipettes, needles and watchmaker's tweezers, they emptied the egg of DNA and transferred into it the nucleus of a frog cell taken from an early embryo. These delicate manipulations paid off in November 1951 when they successfully reconstructed frog embryos that grew into tadpoles. Nuclear transfer 'may have other uses', they speculated in the *Proceedings of the National Academy of Sciences* in March 1952. This failed to anticipate all the hype, activity and headlines that would follow the arrival of Dolly. Nevertheless, their anodyne statement was enough to electrify the biologists of the day.

Briggs and King examined the question raised by Spemann:

could an adult cell retain the ability to direct embryonic development? They found that about half the nuclei from blastocysts generated normal tadpoles, showing that their method worked in principle. But, even though it was no more technically demanding, attempts to use nuclei from the more differentiated cells in the next stage of embryonic development from the blastocyst, a double-walled stage of the embryo called the gastrula, did not work. (One reason is that a fertilised egg divides rapidly compared with the cell cycle of the transplanted nucleus, which winds down as it differentiates.) They concluded that developmental potential diminishes as cells differentiate, and that it was impossible to produce a clone from the nucleus of an adult cell. The implications for any proposal to clone from an adult cell are clear: Dolly was, by the understanding of the 1950s, a fiction, a mirage, an impossibility.

Evidence that every cell of an adult contains the entire recipe to make an individual – the genome – was obtained in the late 1950s by John Gurdon, a graduate student at Oxford University working under the supervision of Michael Fischberg, a Swiss developmental biologist. That Gurdon even began a career in biology at all is remarkable. As a pupil at Eton he had been told that science was not his strong point. His schoolmaster delivered a withering verdict on Gurdon's hope of becoming a scientist: 'This is quite ridiculous. If he can't learn simple biological facts he would have no chance of doing the work of a specialist, and it would be a sheer waste of time, both for him and for those who have to teach him.' Despite this devastating judgement, the time-waster found himself, as a mere graduate student, challenging the wisdom of Briggs and King, by then established figures in the field.

Instead of leopard frogs, Gurdon was trying to clone the African clawed frog *Xenopus laevis* (*Xenopus* is Latin for 'peculiar foot', an apt description of the enormous webbed, five-toed, three-clawed rear feet; *laevis* means 'smooth'). *Xenopus* eggs are about one-tenth the volume of those of leopard frogs and have elastic membranes that make them even harder to work with. But they were easy to persuade to make eggs – they had to be injected late the night before the eggs were needed (this research was not good for Gurdon's social life). And they were plentiful in the 1950s because female *Xenopus* frogs were used in a hospital pregnancy test in which a female *Xenopus* would make eggs if injected with the urine from a pregnant woman. The scientific question Gurdon asked of these frogs when conducting thousands of transplants of cell nuclei was whether the nuclei from these differentiated cells were fully potent – totipotent – and thus able to reconstruct an entire organism.

Gurdon, now Sir John Gurdon in recognition of his outstanding contribution to science, showed that this was indeed the case. He carried out his experiments on tadpole intestinal cells, including some that were differentiated. In 1966 he reported that when he inserted the DNA of a cell nucleus from a juvenile animal into enucleated eggs, he could make a frog every so often. 'In all, 120 transplanted intestinal nuclei yielded seven adult frogs, five of which were fertile,' Gurdon later wrote. 'That experiment was perhaps the single most important one we performed, because it proved that a cell can undergo specialization and yet remain totipotent, retaining all the genetic material needed to make a complete, sexually mature individual. This key experiment justified the view that the cloning of differentiated, and perhaps even adult, cells was at least theoretically possible.'

But questions were raised about these experiments. They went far beyond what the work of Briggs and King suggested should be possible, and the *Xenopus* results were thus greeted with scepticism. Some scientists were concerned that a small but significant percentage of intestinal cells used by Gurdon for cloning might have contained relatively undifferentiated stem cells, because they originated during development in the endoderm, the part of the early embryo that turns into the gut. Others, for reasons that Gurdon himself still does not understand, suspected that germ cells could have been present in the intestine and that these undifferentiated cells enabled his experiments to 'succeed'. In short, people doubted that it was possible to clone with cells that were at a later stage of development than those used by Briggs and King.

These criticisms spurred the Oxford team on. In 1975, Gurdon, Ronald Laskey and Raymond Reeves reported in the *Journal of Embryology and Experimental Morphology* the transplantation of the nuclei of skin cells from the foot webs, providing convincing evidence that Gurdon really was using fully differentiated donor cells. They used a marker, an antibody, to convince the critics that the donor cells were indeed skin cells. Around this time Gurdon also took a spectacular photograph of thirty little albino frogs, cloned by nuclear transfer from the cells of an albino tadpole into the eggs of normally pigmented frogs. The image appeared in the newspapers. Some readers must have been aware of the echoes of Aldous Huxley's *Brave New World* in which 'Bokanovsky's Process' could turn each fertilised human egg into ninety-six perfectly formed embryos ('But alas,' the Director shook his head, 'we can't bokanovskify indefinitely.')

At this point you may be wondering how these experiments fell short of the success that the Roslin team later claimed with Dolly. Perhaps Gurdon's studies lacked the glamour of a mammal experiment. Perhaps, if Gurdon had called his first albino clones Whitey, Footy or even Webster, he could have attracted enough publicity to rewrite cloning history. But more important, his experiments were never able to generate an adult animal from the nuclei of cells derived from an adult animal. He had produced fertile adult frogs from the nuclei of tadpole intestinal cells, but those cells came from a juvenile rather than an adult animal. Although he had grown heartbeat-stage tadpoles from the nuclei of adult skin cells, none of his skin cell embryos survived to become adult frogs. From all his studies, he concluded that transferred nuclei from cells that are more specialised are less likely to support development beyond the early stages than are less specialised cells. Although he did not clone an adult frog from an adult frog cell, his important work did emphasise that the genome remains intact during the course of cell differentiation – a key principle – and nicely complemented what was to come with Dolly.

Today, Gurdon comes across as an archetypal old-school Englishman. Well-spoken, with floppy brown hair and a love of fast cars, he is a man of independent means, blessed with an indifference to mortgages and other tiresome features of everyday financial survival. He has an acute mind and the ability to ask the right questions. Above all else, he remains passionate about his science. Despite his age and his eminence, Gurdon can still be found working at the laboratory bench in the research institute in Cambridge that now bears his name.

Nuclear Rabbits and Mice

The story of mammalian cloning can be traced back to one of Gurdon's pupils, Chris Graham, a chain-smoking intellectual with a slow turn of phrase and a dry sense of humour. Inspired by Gurdon's success with frogs, Graham had moved in the mid-1960s to Oxford University, where he worked in the department of Henry Harris, a Renaissance man. Harris was born and educated 'under a hot Australian sun', first read modern languages and then turned to medicine, though he remained interested in history and writing, whether it was fiction or a vivid account of the science of the cell. Harris had come up with a major insight into cancer after he had fused cancer cells with normal cells and found that the resulting hybrids grew normally – evidence of what are now called tumour suppressor genes, which help to prevent cancer. Using Harris's method of fusing cells with the non-infectious Sendai virus,

Graham wanted to develop a gentle nuclear transfer technique for mice, one that did not involve breaking cell membranes.

The virus is able to link together the outer membranes that protect cells for a short while, like the skins of merging soap bubbles. With two cells nestling side by side, the membranes flow together and fuse to enclose the two nuclei and allow the creation of hybrids. In 1965, Harris and John Watkins had published a paper in *Nature* describing how they had even fused human and mouse cells to form 'heterokaryons' with genetic material from both species. This technique opened up limitless possibilities for the study of the relationship between the nucleus of a cell and the cytoplasm, the signals sent within cells and the function of genes. The media were, of course, more intrigued by the possibilities of man–mouse hybrids – the stuff of nightmares.

Since then scientists have indeed used the virus to fuse unlikely couples such as insects and mammals for basic research. The virus also seemed to offer Graham a way of easing the nucleus of a cell into an egg without the trauma imposed by passing it through an injection needle. Injection threatened to disrupt a mouse egg, which is only one-seventeenth the diameter of a frog egg and just one five-thousandth of the volume. Working at Oxford's Dunn School of Pathology, Graham used the virus to pass a variety of cell nuclei into the one-cell stage (zygote) and two-cell stage embryos of mice. However, he found no evidence that the implanted nucleus participated in the development of any embryo. In fact, even if the new nucleus had taken part in development, there would not have been any live births: the result of these experiments was always genetically abnormal, with twice the usual number of chromosomes. Nonetheless, Graham felt that he was on the path to mammalian cloning.

Today, after a long career that included important work on the genetic alteration of animals, Graham recalls how he wrote 'an outrageous paper' for a 1969 symposium held in Philadelphia in which he indicated that successful nuclear transfer in mammals was just around the corner. Confirming earlier results by the IVF pioneer Robert Edwards, he explained that he could enucleate eggs without damage by using an alkaloid, colchicine, obtained from the autumn crocus, that confuses cell division. He described how the egg could then be parthenogenetically activated (tricked into dividing without the help of a sperm) and, referring to his earlier efforts to fuse cells with the help of the Sendai virus, suggested that successful mammalian nuclear transfer was only a matter of blending these methods in one experiment. This happy event escaped him, partly because he moved the eggs and cells together with a hand-held pipette, not with a micromanipulator, and partly because scientists were still developing reliable methods for keeping a single-cell embryo alive after it has suffered this kind of physical abuse.

Derek Bromhall, another of Oxford's graduate students, would build on the work of Gurdon, Graham, and others. Fascinated by the avenues that would be opened up by transferring Gurdon's frog work to mammals, Bromhall wanted to give up his career as director of fisheries research in Hong Kong to investigate the possibility of nuclear transplantation in mammals. The government chief scientist, Solly Zuckerman, had visited his lab in Hong Kong and backed Bromhall's idea, even though many were doubtful that the feat was possible, given that mammalian eggs are much smaller than those of a frog – in fact, smaller than this full stop.

Bromhall moved to Oxford in 1967 to begin his work on rabbits. Although their eggs are larger than mouse eggs, rabbits are more prone to infection, and surgery must be done in sterile conditions. Over the next seven years, Bromhall developed the microsurgical equipment and culture techniques required for nuclear transfer, attempting to synchronise the cell cycles of the egg and the donor nucleus – by incubating the donor in an atmosphere of nitrous oxide under pressure to arrest the cell cycle – and enucleating eggs by irradiating them with gamma rays or ultraviolet radiation to destroy the DNA (the microsurgery of the day destroyed the eggs). He ran parallel experiments with the Sendai virus and the microinjection of a variety of cells, but concentrated on transferring nuclei from early rabbit embryos (at what embryologists call the morula stage, the morula being the ball of cells that forms before the blastocyst) to unfertilised eggs by microinjection. At the time there were no rabbit antibody markers, as there were in Gurdon's frogs, and it was essential to prove that the cells of an embryo created by nuclear transfer were from the transplanted nucleus – otherwise it could be argued that the embryo had developed parthenogenetically, from an unfertilised egg, which is possible for a few cell divisions.

In an attempt to follow the fate of the transplanted nucleus, Bromhall labelled the donors with radioactive thymidine (thymidine is involved in the synthesis of DNA and in the preservation and transfer of genetic information). Unfortunately, this technique necessitated killing any embryos created by nuclear transfer by slicing them into thin sections which were then coated with a photographic emulsion. Exposed for several days and then developed, the nuclei in the embryos' cells derived from the transplanted nucleus could be identified by clusters of black spots caused by the radioactivity in the thymidine.

The next step would have been to transfer the cloned embryos to the uterus of a recipient rabbit. It was at this stage that the last of Bromhall's research grants, from the Cancer Research Campaign, ran out, and he changed career again, becoming a producer of documentary films. I doubt that he would have been successful in his cloning work. However, at that time very few of the enabling techniques that we took for granted in the creation of Dolly were well established, so when Bromhall reported this work in 1975 it was noteworthy enough to grace the pages of the journal *Nature* on Christmas Day. This was a milestone, the first published attempt at cloning a mammal by nuclear transplantation.

During Bromhall's thesis work on cloning, Gurdon had left Oxford. But the reputation of the lab still attracted visits by key figures in this developing field, such as Andrzej Tarkowski of Warsaw University. Tarkowski had used electricity to persuade mouse eggs to divide, had studied the detailed effects of the egg on an alien nucleus, and later went on to use electricity to fuse mouse embryos. Bromhall's work was 'not very convincing', according to Tarkowski, who today is a professor in the classic central European style – old-fashioned, formal and hierarchical, a serious man with a passion for biology and a talent for photography. Tarkowski thought that one of his colleagues, Jacek Modliński, had provided stronger evidence that nuclear transfer worked in mammals when, in 1978, he injected embryonic nuclei into non-enucleated mouse zygotes and witnessed the development of chromosomally abnormal embryos to the blastocyst stage. 'He presented undisputable proof that the injected nucleus was incorporated into the genome of the operated embryo,' recalled Tarkowski.

While he was trying to clone his rabbits, Bromhall was also visited by Davor Solter, a scientist born in Croatia. Then in his mid-thirties and based at the Wistar Institute in Philadelphia, Solter wanted to find out more about the possibilities of mammalian cloning. Although Bromhall's work was pioneering, Solter was struck by what a struggle it was to get his methods to work. 'I watched Derek Bromhall work for a week without a single successful transfer,' he recalls. There was still much left to be done – but not in Oxford. This early powerhouse of nuclear transfer would become more preoccupied with molecular biology. A decade later, as mammalian cloning moved to other laboratories, Solter became a significant player in the field.

In His Image

These early cloning efforts inspired David Rorvik's book *In His Image: The Cloning of a Man*. Originally billed as the true story of the cloning of a mysterious millionaire, the novel was published in 1978, the year in which Louise Brown, the first test-tube baby, was born. With a readership intrigued by the possibilities raised by that momentous birth, his book caused a sensation, generating many column inches of comment in newspapers, a congressional hearing and endless angst.

This is how Rorvik's story goes. A self-educated, ageing million-aire businessman named Max yearns for a son. He asks Rorvik to use his connections to recruit scientific talent for a secret project to clone an heir from his own DNA. In return he will allow Rorvik to write about the reproductive wonder in the making. According to Rorvik, a team is assembled, headed by a brilliant but overweight scientist code-named Darwin. They are flown to a laboratory on a secret island somewhere in the Pacific. After years of experimenting, they succeed in achieving a pregnancy in a surrogate mother, a local teenager code-named Sparrow. Nine months later, Sparrow gives birth to the first human clone. Rorvik wrote, 'It was not, I thought, exactly the nuclear family. But it was a thrilling sight, this old man, this young girl, this strange baby.' He insisted that his work was non-fiction, but the book was printed with a disclaimer which told readers to decide for themselves. Rorvik refused to divulge details 'to protect the child from harmful publicity'. It is touching to see a journalist express such concerns.

Scientists cried foul, pointing out that, among other things, no one had ever grown a frog to adulthood, let alone a mammal, from the specialised cells of an adult. But the book became a best-seller, and politicians took notice. Briggs had testified at the 1978 House of Representatives hearing on Rorvik's sensational book that 'cloning in man or any other animal is not just a technical problem to be solved soon but may, in fact, never occur'. Briggs and King had their own experiments to draw upon as evidence of this asser-tion. But perhaps they had done something wrong. As they later wrote, 'at the time we could not eliminate the possibility that the developmental deficiencies might be the result of nuclear damage during the operation'.

To give his work a veneer of respectability, Rorvik had cited Derek Bromhall's efforts to clone rabbits. This backfired. Bromhall was infuriated. He remembered that Rorvik had written to him some time before, explaining how he was writing a review of nuclear transplantation, and, 'assuming that Rorvik was an honest researcher', he had sent him the abstract of his doctoral thesis. Bromhall would sue the publishers, J. B. Lippincott, for dragging his name into the furore. He had never consented to the inclusion of his name or research techniques in the book and did not want to be associated with the idea of advocating human cloning. In 1979 a judge ruled that *In His Image* was a hoax. Three years later, the publisher admitted that the book was fiction.

Bromhall himself was no stranger to fictional cloning. He was the scientific adviser for *The Boys from Brazil*, which linked cloning with totalitarianism. Duplicates of Adolf Hitler starred in this 1978 movie version of the Ira Levin novel of the same name. It is an unfamiliar take on the familiar script about Nazis trying to rule the world. Hitler's mass executioner, the geneticist Josef Mengele, having escaped to Brazil, clones ninety-four boys from Hitler's blood and places them with adoptive parents. The boys are, of course, little rotters. But even Nazis know that having the 'right' genes is not enough to grow an authentic Adolf. After all, the real thing might have painted landscapes for the rest of his life had he not been rejected by the Vienna Academy of Fine Arts. The clones are placed in adoptive homes in North America and Europe and raised in a way that roughly resembled the Führer's own upbringing. As each proto-Hitler nears the age of fourteen, his father dies, as did Hitler's own father at that age, thus recreating the microculture for the clones that would allow each mini-Hitler to become evil incarnate. The film tapped into the public's deep unease about perverting science for political and ideological ends.

Intriguingly, Bromhall shot a sequence for the film illustrating the techniques of nuclear transplantation to explain how the cloned Hitlers could have been created. He went on to produce wildlife and medical documentaries, for one of which, following the growth of a baby from conception to birth, he was nominated for an Oscar. With the cooperation of the staff at the Roslin, Bromhall also documented the story of Dolly until the birth of her first lamb, Bonny.

Brave New World

When nuclear transfer began to be conducted on amphibians, and later applied to mammals, eminent scientists were also inspired to think through the implications. Take, for example, the great geneticist J. B. S. Haldane, who helped to create the cultural backdrop to cloning. In his 1924 book *Daedalus* he explores the supposed advantages of 'ectogenesis', the growing of human embryos in artificial wombs. In this way Haldane influenced Aldous Huxley's 1932 novel *Brave New World*, in which lower castes in a dystopian society are cloned to serve totalitarian ends.

Three decades later, in 1963, Haldane delivered a remarkable speech entitled 'Biological possibilities for the human species in the next ten thousand years' in which he introduced the term 'clone' to a wider audience. His views were born of a naive confidence and optimism about science that was prevalent at that time. Oblivious of some of the darker implications, they were driven by excitement at the thought of cloning a Mozart, rather than cloning an embryo of unknown potential (let alone one of the hoi polloi). Haldane told his audience:

Perhaps the first step will be the production of a human clone from a single fertilised egg, as in *Brave New World*. But this would be of little social value. The production of a clone from cells of persons of attested ability would be a very different matter, and might raise the possibilities of human achievement dramatically. For exceptional people commonly have unhappy childhoods, as their parents, teachers and contemporaries try to force them to conform to ordinary standards. Many are permanently deformed by the traumatic experiences of their childhoods. Probably a great mathematician, poet or painter could most usefully spend his life from fifty-five years on educating his or her own clonal offspring so that they avoided at least some of the frustrations of their original.

On the general principle that men will make all possible mistakes before choosing the right path, we shall no doubt clone the wrong people. However, everyone selected for this purpose will presumably exceed the median considerably in some respect, if only as a humbug. And the greatest humbugs, like Hitler, would hardly relish the thought of producing a dozen possible successors with their own abilities, and youth to boot. Possibly a movie star at the age of forty might have similar feelings.

Assuming that cloning is possible, I expect that most clones would be made from people aged at least fifty, except for athletes or dancers, who would be cloned younger. They would be made from people who were held to have excelled in a socially acceptable accomplishment. Sometimes this would be found to be due to accident. The clonal progeny of Arthur Rimbaud, if given favourable conditions, might have shown no propensity for poetry, and become second-rate empire builders. Presumably such a clone would not be further grown. Other clones would be the asexual progeny of people with very rare capacities, whose value was problematic, for example permanent dark adaptation, lack of the pain sense, and special capacities for visceral perception and control. Centenarians, if reasonably healthy, would generally be cloned, if this is possible; not that longevity is necessarily desirable, but that data on its desirability are needed. Centenarians who could continue to learn systematically up to the age of thirty would almost certainly be useful, and probably happy members of society.

Haldane, who died the next year, was not the only scientist to warm to the possibilities of cloning during this age of wide-eyed optimism. The Nobel laureate Joshua Lederberg also discussed how to use cloning to 'improve' humankind in an article in the *American Naturalist* in 1966. Others were not so sure. Four years later came Alvin Toffler's influential warning about the disorienting pace of scientific and technological change, *Future Shock*, which, like Haldane, anticipated the story line in *The Boys from Brazil*. Toffler mused on how there was a certain charm to Albert Einstein's bequeathing copies of himself to humanity, 'but what of Adolf Hitler?'

Haldane's sister Naomi Mitchison further underlined the naivety of the initial discussions of cloning in her 1975 novel *Solution Three*. Her title was meant to echo the Nazis' Final Solution, the drive for racial purity that led to the horrors of the death camps and human experimentation. This science-fiction novel had a more compelling provenance than most: it was dedicated to the DNA pioneer Jim Watson, who first suggested the idea for the book. Another key influence was embryologist Anne McLaren (who, incidentally, had appeared as 'The Child' in Alexander Korda's 1936 sci-fi film *Things to Come*). And Mitchison's childhood friends included Julian Huxley, who would become an eminent zoologist, along with his younger brother, Aldous.

Mitchison populates her novel with lesbian, gay and straight characters who live in an emancipatory society based on mandatory homosexuality, banned heterosexuality, and a new model for reproduction using 'clone mums' and systematic conditioning of their cloned children. *Solution Three* is driven by the need to cope with rising world population and falling food production, as efforts to create a narrow range of superplants have left major crops vulnerable to disease. Mitchison explores a feminist agenda to use cloning on human guinea pigs to wipe out sexism, racism, war and other trappings of an oppressive, diverse and patriarchal society to produce instead a uniform, predictable and peaceable society, a forest of human genetic excellence. Her novel reminds us about the importance of nurture, not just nature, about the value of diversity and about the unexpected costs of even the best-intentioned technological fix. She ends on the optimistic note that flexibility, not dogma, can win out.

Today her older brother's speech fascinates and chills in equal measure. It reminds me of one of the first things I learned at university – that there is no clear relationship between intelligence and a sense of social responsibility. Because of this I quite see why the media still use *Brave New World* as shorthand for the direction in which the scientific control of reproduction could take us without thinking through the implications and subjecting them to proper scrutiny.

A False Start

Although it is very relevant to the cloning of mammals and people, the early work that so inspired Haldane and Lederberg must be set in its proper context. Frogs and salamanders are amphibians. They are easy to work with because their eggs are so much bigger than our own – human eggs are minute in comparison, as ping-pong balls are to pumpkins. In addition to being measurable with a school ruler, the eggs of amphibians are easy to get at, as easy as collecting frog spawn in a jam jar. But mammal eggs never see the light of day. They are hidden deep in the female body in the uterus and oviducts. At the very least, many new techniques would have to be developed if this pioneering work with amphibians was to be applied to mammals. Despite advances made in Oxford, Warsaw and elsewhere, many scientists expected, quite understandably, that the hurdles would be huge, and some feared that it might never be possible to clone mammals at all.

That view was challenged by some truly fantastical research on mice. Mammalian cloning seemed to have been launched in earnest with experiments by the American Peter Hoppe and Karl Illmensee of the University of Geneva, the latter a trim and dashing German embryologist regarded as something of a wunderkind. In the 1970s, at a scientific gathering known as a Gordon Research Conference, they astounded their peers by claiming to have created mice with all-male or all-female parents. In a 1977 paper in the *Proceedings of the National Academy of Sciences*, they described how they had done this – by removing a female or a male pronucleus from an unfertilised egg. They made up for the loss of genetic material by doubling (diploidising) the genetic material in the remaining pronucleus with the help of a fungal metabolite called cytochalasin, which interferes with cell division by inhibiting the replication of the proteins that form the cell's supporting framework, the cytoskeleton. We now know that, because of the influence of a phenomenon called imprinting (described in the next section), this is a dubious proposition and the resulting embryos do not develop beyond mid-pregnancy.

Then, in January 1981, Illmensee and Hoppe announced in the journal *Cell* that they had cloned three mice. These mice were all the more remarkable because a succession of failed attempts had begun to make biologists suspect that cloning a mammal was impossible. Thanks to these dazzling feats of reproductive science, and helped in part by his considerable presence, Illmensee was in great demand at conferences and on the international scientific lecture circuit. In a way he even helped to create Dolly: his lectures would inspire two key figures in her story, Keith Campbell and

Steen Willadsen. Keith attended Illmensee's lecture in 1984 at the University of Sussex, when he was beginning his doctorate. 'Who knows,' Keith later reflected, 'without him I might not have got involved in this field at all.'

But others had problems in replicating the amazing feats of Hoppe and Illmensee – in fact, they have never been repeated. And although anecdotes may inspire and intrigue scientists, they don't sway them. Illmensee was secretive about his methods, and his reputation for having 'golden hands' began to wear thin. Even members of his own group grew frustrated and mutinous. In 1983 matters came to a head when a commission of inquiry accused him of sloppiness and asked him to repeat a series of experiments. Crucially, however, these experiments were not concerned with cloning, and there was inadequate evidence that the pair had fabricated data.

Among the attempts to follow up this work, one stands out because of its impact on future thinking. More than anything, it explains why some scientists greeted Dolly's arrival with gasps of disbelief. Ironically, although – unlike the experiments that inspired it – this one study was reproducible, it held back the field for years.

Well, that is at least according to many popular accounts of the history of cloning that describe an attempt to reproduce what Illmensee and Hoppe had done, with a painstaking programme of experiments at the Wistar Institute of Anatomy and Biology, in Philadelphia. The studies were carried out in the early 1980s by Davor Solter, by this time a respected embryologist and an American citizen, and his student James McGrath. In common with others, they wanted to learn more about differentiation. But, like everyone else (such as Chris Graham in Oxford), they could not get the Illmensee–Hoppe method to work. When Illmensee paid a visit to the Wistar, they asked him to give them a demonstration. For one reason or another, he never did. Perhaps Illmensee and Hoppe had been able to get three cloned mice but unable to repeat their good fortune. The efforts of McGrath and Solter, however, suggested otherwise. Solter himself believes that the 'charitable explanation is that Illmensee and Hoppe mixed up some embryos and thought they had succeeded'.

Their technique, and all the efforts to replicate it, was doomed to fail. Illmensee and Hoppe had carried out nuclear transfer into enucleated zygotes – fertilised eggs – rather than unfertilised eggs, because it was wrongly thought that the former would provide better support for development (though this baffled researchers who did the pioneering work with amphibians, in which unfertilised eggs were used). 'The original experiments have since been repeated many times by successful cloners, and the results were always negative,' said Solter. But that did not make the exercise a waste of time.

In attempting to clone mice in this way, Solter and McGrath had made a useful contribution to nuclear transfer technique that they described in *Science* in 1983 (in a paper first rejected by the rival journal *Nature* as having little practical application). Until then, scientists had not only penetrated the zona pellucida of the egg to carry out nuclear transfer, but pricked the inner – plasma – membrane too. As Solter put it, 'Everyone was killing eggs by jabbing needles into them.' The Wistar team established a gentle way to remove the nucleus from a fertilised mouse egg and then introduce a new nucleus with as little disturbance as possible, leaving the inner membrane of the egg or zygote intact. We drew upon this wisdom in the creation of Dolly. Using this method of nuclear transfer, they created live mice (but not clones, since they could make only one copy of each zygote). 'Weeks after our paper came out, several laboratories started successfully using our technique in attempts to clone large farm animals,' Solter recalled.

He and McGrath went on trying to reproduce Illmensee's work in a tenacious and methodical set of experiments. They had already shown that the simplest form of nuclear transfer – from one newly fertilised egg to another – could succeed, underlining how their technique was capable of delivering results. That was about as far as they could go, however. Almost all transfers from two-cell embryos reached the blastocyst stage, but transfers from four- and eight-cell embryos and from the inner cell mass of more advanced embryos (as used by Illmensee and Hoppe) did not. They had no success with using eggs as recipients either.

Their failure had been so broad and so systematic that their results were published by the journal *Science* in December 1984. The Wistar results were more than negative – they effectively ended Illmensee's research career because of the questions they raised about just how he had obtained his results. Hoppe continued to search for the magic batch of cytochalasin to enable him to create uniparental mice, but it was not to be. As the Wistar team concluded, 'the cloning of mammals, by simple nuclear transfer, is biologically impossible'. And that was that. Spemann's 'fantastical experiment' remained a fantasy.

From the Ashes

Today, Davor Solter is the director of the prestigious Max Planck Institute of Immunobiology in Germany. His sandy hair is going grey, and he can sometimes look burdened. Despite many years on the international circuit, he still speaks English with a Croatian accent. His colleagues find him a warm, charming, soft-spoken man who is generous in discussing his findings and sharing his insights. He becomes animated when he discusses his passions – films and books (Philip Pullman is a favourite author) – but on certain scientific subjects, such as the history of cloning, he often gets angry.

Solter is weary of how his 1984 *Science* paper has so often been condemned as a hammer-blow to efforts to clone mammals, denting the confidence of people like me and holding back the field for years. The Wistar studies did indeed suggest that the potential of transferred nuclei to support development all but vanished within the first few cell divisions of an embryo. Solter and McGrath suspected that it had something to do with how the genes in mice go into action so early in development compared with amphibians that they could not be reprogrammed by transfer to an enucleated zygote. In *Science* they suggested that the key to the problem might be found in earlier studies they had conducted at the Wistar, as had Azim Surani of the Institute of Animal Physiology in Cambridge, England, in which mouse embryos that had two fathers or two mothers (created by exchanging maternal and paternal chromosomes) all died. The implication is that the embryos needed genes from each parent.

We now realise that with this work Solter and McGrath had made a key contribution. This phenomenon, called imprinting, ensures that, for certain genes, only the copies from the mother or father are used during development. This, too, would make reprogramming harder, so they concluded in *Science* that their results *suggest* that the cloning of mammals by simple nuclear transfer is impossible. Given what we know today, their thinking was far from wrong. But they used zygotes rather than eggs, an approach that we now know was less likely to be successful. The Wistar team were also unwise to use the word 'impossible' and have paid for ending on this wrong note ever since.

But perhaps what rankles Solter most of all is his frequent portrayal as the powerful scientist whose devastating judgement retarded research into cloning. He protests that this is laughable, given his position in the field in 1984. 'People were already doing nuclear transfer in farm animals with promising results for the past year, and nobody that I spoke with at the time was in the least dismayed with our results.' Several teams, including my own, were indeed pressing on, not least because there are hundreds of factors that affect nuclear transfer, and although he had been diligent and thorough, Solter had not explored all the possibilities.

We were mere animal scientists who worked in fields, barnyards and slaughterhouses. To us there was practical value in being able to make several copies of embryos recovered from the very best farm animals. Some registered elite animals have values of around £500,000 each, and their offspring sell for up to £100,000. Potential profits provided the stimulus for us all to press on. Unknown to us at the time was the fact that nuclear transfer is easier to carry out in many farm animals than in that ubiquitous laboratory animal, the mouse. Researchers who were developing methods to clone farm animals included Neal First of the University of Wisconsin, where Randy Prather, Frank Barnes, Mark Westhusin and James Robl, among others, were working on how to clone prize cattle embryos with the backing of the W. R. Grace Corporation.

Moreover, the gloom of Solter and McGrath did not put off Steen Willadsen, an inventive, extrovert and energetic Dane whose work was recognised with the Pioneer Award of the International Embryo Transfer Society. I am sometimes called the 'father' of Dolly – which is unfair, given the huge contributions from Keith Campbell and others at the Roslin – and somewhat daft too. Well, this brash and mysterious Dane is without doubt her 'mother'. Willadsen began as a veterinarian, then worked as a research scientist, first in Britain and later in the United States, making many important advances. Now working on human reproduction, he is a brilliant maverick who does not suffer fools gladly.

Willadsen had also worked under Chris Polge at the Agricultural Research Council Unit of Reproductive Physiology and Biochemistry in Cambridge, where I had frozen boar sperm and produced the first calf from a frozen embryo. Just before he arrived, I had departed for the Roslin Institute. Willadsen inherited my fellowship and my research project in an environment of great academic freedom, where good ideas and innovation could flourish without interference. And this freedom allowed him to follow the apparently discredited field that was mammalian cloning.

Willadsen played around with cloning by embryo splitting. One problem is that when the protective zona pellucida is damaged, the mother's system destroys the embryo up to the blastocyst stage. Willadsen overcame this by dividing a sheep or cow embryo in two, then encasing the halves in a couple of millimetres of protective agar, a jelly made from seaweed. That way, the embryos could be allowed to develop in the reproductive tract of another female before recovery – so that the agar could be picked off with a hypodermic needle once it had done its job – and then implanted in a surrogate in a sufficiently advanced stage to survive until term. As well as making it easier to clone by embryo splitting, this technique made it much easier to manipulate embryos, expanding the horizons of reproductive science. With Carole Fehilly, who would become his wife, Willadsen even mixed cells from embryos of different species to create a geep (goat–sheep).

The geep was not the first artificial chimera. That honour goes to a mouse born on 6 March 1961 as a result of an experiment by Andrzej Tarkowski, then in the Department of Zoology at University College of North Wales, where he worked as a Rockefeller Foundation Fellow. Almost four decades later, in a retrospective piece in the *International Journal of Developmental Biology*, the Polish embryologist cheerfully referred to this 'crazy' project. Back in the early 1960s he had found that you could combine two mouse embryos consisting of a few cells and end up with a single animal whose salt-and-pepper fur was the only visible indication of its dual origin. At the same time, and to his great surprise, Tarkowski discovered that another scientist was doing exactly the same crazy experiment – Beatrice Mintz of the Fox Chase Cancer Center in Philadelphia. Like Driesch's flexible sea urchin embryos, chimeras underlined how the cells in an embryo responded to their environment. Another way to make them, by injecting cells into blastocysts, a method devised by Richard Gardner and Ralph Brinster, would open up even more possibilities for introducing foreign cells into the embryo.

Willadsen and Fehilly were the first to mix cells from different species. This apparently bizarre experiment to create a geep chimera had a serious purpose. At the time, they were interested in determining how a mother tolerates the presence of the foetus during pregnancy, an immunological issue that sheds light on what can go wrong and lead to miscarriage. Although a sheep will not tolerate a goat embryo and vice versa, both will tolerate a geep embryo. Furthermore, some cells in a chimera embryo form the foetus, while others form the placenta. This explains another rationale for Willadsen's research: this approach could allow him to create a chimera in such a way that a mother from a common species could give birth to a foetus from an endangered species: you could make the chimera so that the placenta forms from one set of tissues (of the common species), and the foetal tissue from the other (of the endangered one).

Willadsen pressed on with his cloning research. In 1986 he stunned everyone by announcing in a paper in *Nature* that he had cloned sheep from early embryos. This work is now recognised as a critical milestone in reproductive science. Because Illmensee's feats had not been verified, and McGrath and Solter did not assess development all the way to birth, Willadsen's lambs were the first mammals of any kind to be cloned – beyond any doubt – by nuclear transfer. He had, of course, tried to achieve the feat using Illmensee's methods, but like everyone else he failed.

Undeterred, Willadsen overcame the problems by sheer grit. When I once visited him in Canada, I was amazed to find that he did much of his research solo. He made up for the lack of helpers by working hard. Cloning requires many different skills, and usually they are provided by a big team, as will become clear in the next chapter. Yet in this laboratory he often did all of this work on his own, even down to the anaesthesia and surgery on the same animal. He also had boundless ingenuity. When Solter and McGrath tried cloning, they had used single-cell embryos as recipients for their donor nuclei. But Willadsen reasoned, rightly, that an unfertilised egg would be better because it was brimming with machinery – proteins – to act on the DNA from sperm, and this machinery could reprogram the DNA in a donor nucleus. All the work on amphibians suggested that this was the way to go.

Willadsen looks back to when he wrote the 1986 *Nature* paper on cloned sheep:

> I made sure that my 'predictions' were already realised by the time I let *Nature* publish them. By then I had not only cloned cow embryos, but also integrated embryo sexing and embryo freezing into the procedure, and I had demonstrated that serial nuclear transplantation worked in principle. In other words, I had established a framework within which one could hope to tackle all reasonable cloning needs of cattle breeders – which was the official objective I had stated to the [Agricultural Research Council].
>
> Before hardly anyone knew what was happening, a nuclear transplantation procedure of general usefulness had been developed (1983–84) and both sheep and cow embryos had been successfully and repeatably cloned [by which] I mean that lambs were born (August 1984) and calves were under way (November 1985). And insofar as this could be done by one man, I did it.

Bravo! To achieve that as a member of a team is impressive. To do it alone, and create two lambs from three eggs, is heroic.

There was more. Working at a Texan ranch for Granada Genetics, a company that eventually fell by the wayside when a tax break was ended, Willadsen pushed back the boundaries once again. Some believed that it was not possible to use visibly differentiated cells for nuclear transfer, but Willadsen had other ideas: ever the pragmatist, he would obey this 'law' of biology only after he had probed and tested it for himself. All that bothered him was that it became harder to obtain these cells because they clung together more tightly in more advanced embryos.

Going against the grain of dogma, Willadsen had by 1987 produced live calves by nuclear transfer from embryos that had progressed to the 32- to 128-cell stages, but this work was not presented in public for several years. At this stage of development the cells of the embryo can still be readily separated. The nucleus from one of these cells was then transferred into an aged egg. One might expect a fresh egg to be better. Yet the eggs Willadsen used were older than those used for *in vitro* fertilisation, for example, which rather shook me at the time. With hindsight it is possible to see that Willadsen had, by trial and error, chanced upon one of the secrets of the Roslin team's later success. By using old eggs, he had found one of the effective ways of coordinating the growth of the donor cell nucleus and the recipient egg. (Later research revealed why this is important. An aged egg makes it easier to activate the development of the clone. In our later research, however, we discovered that old eggs do have one disadvantage: they are not so good at reprogramming a transferred nucleus and so would struggle to reprogram the DNA in an adult cell.)

Willadsen's success was remarkable. By this stage of development, the early embryonic cells had differentiated into the first two cell types: the outer trophectoderm, the shell of cells that becomes ancillary equipment such as placenta, and the inner cell mass, which is destined to be the embryo proper. His research suggested for the first time that nuclear transfer in mammals was possible from at least these early embryo stages. Willadsen's work would provide much of the impetus for Dolly.

That same year, 1987, also saw Illmensee's controversial feat, the one that originally inspired Willadsen, at last come to fruition in a convincing way. Yukio Tsunoda and colleagues at Kinki University announced that they had cloned mice after transfer of nuclei between very early embryos. Illmensee claimed that this vindicated his work. Crucially, however, Tsunoda had used a different approach. As Solter dryly remarked at the time, 'The Moon landing by the Apollo mission did not prove that the techniques and results described by Jules Verne actually worked.'

Willadsen's Inspiration

My path to Dolly began with a stroke of serendipity while drinking with colleagues in a Dublin pub in January 1987. I was attending my favourite conference, the annual meeting of the International Embryo Transfer Society. In those forlorn days of relentless cutbacks by the British government, I was lucky to be there at all. I was the guest of my director, Roger Land, who paid for my air fare with money he'd earned from some consultancy work. Despite our many differences, I owe Roger a great debt for giving me that opportunity. In the dimly lit pub I was told of pivotal and illuminating new work by Willadsen, now working in the heart of Texas, by one of his colleagues, the veterinarian Geoff Mahon (whom I thought of as Willadsen's 'minder').

Often the most valuable parts of scientific meetings are these informal chats, when fellow researchers gossip and sometimes share ideas or new results that have yet to be published in formal papers. (Sadly, the exchange of ideas has diminished as research has become more commercial and there is more pressure to patent.) From Geoff I learned that Willadsen had used nuclear transfer with a bovine embryo that was a few days old, at the early blastocyst stage, when they had around sixty-four cells that could still be separated. Of course, this experiment, for Granada Genetics at College Station, Texas, could not possibly have worked, according to the dogma of the day, because the genes of the cattle embryo had already gone into action and its cells had started to differentiate. But Geoff told me that Willadsen had succeeded. Various pennies dropped and bells clanged (though I am not the sort of person who runs naked down the street shouting 'Eureka!').

As Roger Land and I flew back over the Irish Sea on a small and rather noisy plane, I enthused about what I wanted to do. The very fact that Willadsen had produced calves from blastocysts almost certainly meant that he had carried out cloning with the stem cells of the inner cell mass, the part of the blastocyst that mostly turns into the embryo. Working in Oxford, Rosa Beddington and Liz Robertson had already shown that mouse embryonic stem cells retain many of the characteristics of the cells from the inner cell mass. Now I could see how to make genetic modification easier. Experience with mice was showing that precise genetic change was more efficient in embryonic stem cells than in differentiated, adult cells. So if we could derive embryonic stem cells from livestock, we would for the first time be able to effect genetic change in the embryonic stem cells and, when it worked properly, create a clone from one of those GM cells. Together, the techniques of deriving and modifying stem cells and transferring their nuclei from them would revolutionise genetics in livestock. As well as being less hit and miss than earlier methods, this approach would not rely on so many embryos and would be much quicker.

I managed to visit Willadsen, who was by then working in Calgary, Canada. He was not only fun and hospitable, but also a gentleman (when scientists say that someone is a gentleman, or generous, they mean that he openly discusses his ideas and reveals those apparently insignificant details of experimental method that are crucial for success). We walked and chatted in the scrubland near his laboratory, Willadsen carrying his young child with him. He gave me plenty of useful advice. Later he showed me the technical details and even demonstrated how to protect newly constructed embryos with a coat of agar, forming shells that vary from only a fraction of a millimetre to a couple of millimetres across.

But there was one problem. No one had yet isolated embryonic stem cells from sheep. Even as I write this, no one has pulled off this trick. As it turned out, that would not matter because of a twist in this story, one that led to Dolly. With my colleagues I would find that, amazingly, you could do what Willadsen had done but with adult cells – cells which supposedly were already committed to being skin, blood cell, or whatever. We could turn back the developmental clock. We could rejuvenate cells. We did not have to grow sheep embryonic stem cells in order to genetically alter sheep.

Megan and Morag

It is worth standing back for a moment to emphasise two things about the work that led to Dolly. First, after years of participation in pioneering projects to add genes to livestock, we at the Roslin were all too aware of the limitations of early techniques for altering animals genetically. They were inefficient and unreliable, and worst of all there was no way of deleting harmful genes or of implanting new ones in a precise way. Our hope, as we sought funding for the next stage of our research, was to achieve two new things: to derive embryonic stem cells from livestock, and transfer nuclei from them. If we could do that, we could carry out genetic surgery on those nuclei. Mouse geneticists were by then accustomed to being able to make, or 'target', precise genetic changes through the use of embryonic stem cells. This technique had revolutionised molecular genetics in the mouse. We wanted to be able to do the same with breeding farm animals so as to make them hardier, faster-growing and disease resistant.

Second, our achievement rested on hard work by a lot of other people. The group at the Roslin are sometimes credited (wrongly) with the development of nuclear transfer in mammals. As is now clear, that credit is due mostly to Willadsen. And, of course, the origins of his work stretch back even further. Indeed, I did not even introduce the nuclear transfer technique to the Roslin Institute. I had been interested in the development of the embryo in the womb, and in particular why so many embryos perish even in healthy females and what the role of maternal hormones is. In 1985, a Brazilian vet from a British family, Lawrence Smith, came to work for me at the Roslin to do experiments on the egg in the oviduct. After a few weeks he decided that this was boring and that he wanted to try his hand at nuclear transfer instead – a wise choice. By then, Willadsen was blazing the trail.

It was Lawrence Smith who taught me the basics of cloning. Together we produced Roslin's first cloned lambs – four in all – by nuclear transfer, in 1989. As well as helping to confirm Steen Willadsen's findings, that work marked the Roslin as a centre for cloning. Lawrence also taught Bill Ritchie, who carried out the nuclear transfers in our later, critical experiments that led to the birth of Dolly. As the media like to tell us, the principle of cloning is extremely simple. You need two cells in order to produce a clone. One is an egg recovered from a donor ewe at about the time when normally it would be mated. The other is an adult cell from the sheep you want to clone. First you remove the genetic information from the egg, then you introduce genetic information from the adult cell by fusing the two together. This 'reconstructed egg' is grown for a few hours or days in the laboratory, and is then transferred into a surrogate mother and given the chance to develop into a lamb. Five months after transferring the embryo to a recipient ewe, one can expect to see a cloned lamb – if one is very lucky. Easy. (More of the grind, the slog and the sweat behind the feat later.)

Crucially, Lawrence had shown that one important ingredient of success was the state of the nucleus of the donor cell – that is, whether it had just divided or was still preparing to divide. He completed an exquisite analysis of the effect of varying the combination of early and late stages in mouse nuclear transfer. This insight was extended when, in 1991, the Roslin's nuclear transfer team was joined by a long-haired chap from Birmingham with an anti-establishment demeanour and, to complete the stereotype, a love of sinking pints in the pub. Our new 'postdoctoral higher scientific officer' was Keith Campbell, who was to play a central role in the Dolly project.

The son of a seedsman, Keith did not exactly shine at his grammar school, leaving without A levels. He became a medical technician ('I resigned my job the day I qualified to do it'), took a job in a pathology lab in Yemen, returned to Sussex to help fight Dutch elm disease, which was devastating Britain's elm trees, and then worked on cancer. When we met him, Keith had an arty look about him and, as Colin Tudge, with whom I wrote *The Second Creation*, once put it, 'something of the mien of a folk singer'.

As he talked over lunch with me and my daughter (Naomi was spending time at the Roslin for work experience), we both were struck by the clarity of his ideas. Keith recalls it differently: 'I ranted on for about an hour and a half, and they all fell asleep.' The author Fay Weldon memorably described him as being as 'wild-haired as a genius should be. Charismatic, inspiring, an ideas man come to this kind of genetic technology recently, a schoolboy so interested in tadpoles that his mother used to have to sweep baby frogs from the kitchen.'

Keith had the ideal technical background for what I wanted: for his doctoral project at Sussex University, he had studied the cell cycle in *Xenopus*, and was familiar with the work by Gurdon and Illmensee and, much more important, he was fascinated by it. He had also picked up a profound insight into differentiation during his first doctorate, while working for a cancer research charity at the Marie Curie Institute in Surrey, an effort that petered out because his supervisor became ill. Because of this background, he suspected that Spemann's fantastic vision could be realised and adult cells used in nuclear transfer. After all, he reasoned, certain tumours (teratomas) consist of a blend of hair, bone, fat, muscle and other cell types that probably all originated from just one adult cell, or from transdifferentiation, in which one type of differentiated cell turns into another. Indeed, stem cells are now thought to play an important role in cancer because they replenish tissues and organs throughout a person's lifetime, and their progeny are more vulnerable to accumulating enough genetic changes to become cancerous.

The Roslin in turn offered Keith, who had a family, a steady income, employment near where they lived (Keith was by then doing a postdoctoral fellowship at the University of Dundee) and a chance to clone and to apply his ideas about the cell cycle and differentiation. Whereas Keith was driven by basic curiosity alone, I also wanted to use sheep embryonic stem cells for cloning. In this way I hoped to create new opportunities for genetic modification: modify the stem cell and, only when you have that absolutely right, go on to clone and to implant embryos.

Keith focused on the process of nuclear transfer itself. His first critical contribution to the cloning effort at the Roslin was to investigate what it is about the state of the nucleus that primes it for nuclear transfer. He found that the efficiency of nuclear transfer was very limited – unless, that is, the cycles of the donor cell and egg were coordinated. Here it helps to think of a cell as a very exotic chemical clock. Each beat – or division – of that clock is controlled by a complex cycle of chemical reactions. Unsurprisingly, to make cloning work well you have to understand this mechanism. If cloning is attempted on the wrong 'beat', as was done so many times before with more differentiated cells, it does not work at all. Indeed, the better we get to understand the baroque cycle that ticks within our cells millions of times each second, the more amazing the day-to-day business of life itself seems to be. And the early crude attempts at cloning by Spemann, Briggs and others, though of limited success, seem little short of miraculous.

Thanks to the important work by Keith on cow fibroblasts, we would find that for nuclear transfer to succeed we had to use various ways to strictly coordinate the cell cycles of the egg and the donor cell. One approach was to mimic Willadsen's method of using aged eggs. We specifically stimulated the egg so that it would activate before the transfer of the nucleus. Such a preactivated egg is a universal recipient, because it is suitable for a nucleus at any stage of the cell cycle. However, as we now know, the fact that it is aged or activated means that the egg is not best able to reprogram the transferred nucleus – the crucial step that makes it possible to rejuvenate differentiated cells.

Alternatively, and preferably, the egg could be in a particular state, the so-called MII state. Such an egg appears to have the best combination of factors for reprogramming. Appropriately enough, this is the same state that eggs reach when they mature in an ovary ready for fertilisation. MII is shorthand for metaphase II, one of the stages in egg development (meiosis). The 'II' indicates that meiosis has proceeded to the second cell division. 'Metaphase' is the stage in the cell cycle at which the cell has no nucleus and the chromosomes have 'condensed': they lie down its middle, ready to be duplicated and divided. The egg is rich in a substance called meiosis promoting factor (MPF; the M can also stand for maturation or mitosis, depending on which scientist you are talking to), which, as the name suggests, speeds the process along. We were lucky to be using sheep, and so could catch eggs in this receptive stage; rats race through metaphase II, and it is only by the use of drugs to brake their development that they have been successfully cloned.

The next stage is to prepare the MII egg for reprogramming the transferred DNA by emptying the egg of its nucleus, and thus of its DNA. Because these egg cells do not have a nucleus, but rather have free chromosomes, the term 'enucleation' is used loosely for the removal of the nuclear DNA. Crucially, because the nuclear membrane has melted away at this stage, special factors are present, which can bind to DNA to get the process of replication under way.

Just as important as the choice of egg is the control of the cycle in the cell that is to donate DNA to the emptied egg. The cell must be at the beginning of the cell cycle, ready to move on by copying the DNA and growing. This much we had established within a year or two of starting the project. A practical limitation to the use of this new knowledge was that embryonic cells cannot be stopped at the beginning of the cycle. Our Spanish student, Pedro Otaegui, who was doing parallel experiments in mice, spent hours setting up embryos so that they were at an appropriate stage just when he needed them, working night and day, as is often the lot of students.

A colleague, Jim McWhir, had continued the quest for embryonic stem cells in sheep and other livestock that he had begun with one of the pioneers of the field, Martin Evans, in Cambridge. Despite intense efforts he just could not find a way to keep the stem cells from growing for more than a few passages (i.e. through a few cultures). Cells continued to grow, but they were not embryonic stem cells. We agreed that it would be helpful to know when the cultured cells began to lose the flexibility (their totipotency) as they replicated in culture. We could do this by using Keith's methods of nuclear transfer to assess the potential of the cells.

So Keith, Jim and I embarked on experiments to clone from cultured cells that had been through more and more passages, seeking that cut-off point when totipotency evaporated. Using the then standard method, we predicted that we would obtain lambs from the early cultures, but not the later ones, showing that a change had occurred. Success in some of our cloning experiments would lead Jim to believe that he had found the elusive embryonic stem cells (he called them TNT4 cells, meaning totipotent for nuclear transfer). Today, we know that he was mistaken. However, what we had was itself very interesting. These were cells that had only just begun to differentiate, and we found that we could stop them at the beginning of the cycle – we starved the donor cell to send it into a quiescent state. Simply by our removing a source of nutrients for five days, the cell was stopped where we needed it to be.

Using this technique, we cloned Welsh Mountain ewes from the embryo-derived cells that had been cultured in the laboratory. We made each by using a spark of electricity to fuse an embryonic cell with an empty sheep egg. In this way Keith and Bill Ritchie produced 244 embryos, of which 34 developed to the point where they could be implanted. This was not many embryos, but we lacked the money to do any more. Then we held our breath. We knew that if these experiments worked, it meant that we had made embryos from clearly differentiated cells and we would have taken the science of cloning to a new level.

Originally five clones were produced, but three died – two within minutes of birth and the third within ten days, of a hole in the heart. The survivors were born a few days apart to different surrogate mothers in June 1995. The lambs were originally referred to as 5LL2 and 5LL5, where the first 5 denotes 1995, LL is the farm designation and the second number is the order in which the clones were born that year.

The names of these pioneering sheep were changed by Marjorie Ritchie, the Scottish manager of our Large Animal Unit, who had helped to bring them into the world. Marjorie, the veteran of hundreds of embryo transfer experiments, loved the camaraderie of traditional nuclear transfer (as she put it, they were 'happy days but also very busy days') and married my colleague Bill, the cellular surgeon. She named the new arrivals after her own initial letter: Megan to honour the Welsh Mountain ewes that supplied their genes, and Morag to celebrate the Scottish Blackfaces, the ubiquitous local breed that provided the eggs. We did not celebrate their birth, since they were merely two of a series of experimental animals – and, as

Marjorie points out, it is only now that we appreciate how significant they were. They did make headlines when we unveiled them in
1996, but generated nothing like the fuss that was to come with
Dolly.

Breakthrough

Megan and Morag were very important to me. This was our first
groundbreaking experiment. The embryonic cells that had been
used to create them by nuclear transfer were multiplied in culture
until they were nine days old and thus partly differentiated. They
carried molecular markers of differentiation too. These cells were
from embryos significantly more advanced in their development
than the blastocysts used by Steen Willadsen in his critical experiment. The experiment was important because it showed that by
paying attention to the details of the cell cycle, one could reverse
this differentiation and 'rejuvenate' cells.

As if to underline this, Keith became involved in almost every
stage of the process. He moved out of the lab to shave the bellies of
the sheep and disinfected them for operations, helped to gather eggs
and embryos from donors, and even slept in an office next to the
sheep towards the end of their gestation so that he could regularly
check for the first signs of birthing. Once he was tailed by the local
police as he went for a sandwich in the middle of one night ('they
were just as bored as I was'). During the perpetual twilight of the
Scottish summer nights he shared this chore with Bill Ritchie, John
Bracken, and others before rushing home to say goodbye to his
children before they left for school. As it turned out, all the lambs
were born in the afternoon.

Megan and Morag proved that cloning from a cultured cell, one
that was partially differentiated, was possible. They can be thought of
as forming one great clone with the other lambs that were born

alive as a result of the process, together with the foetuses that did not make it to term, the embryos that never made it to implantation and the cells that were used for nuclear transfer. Keith and I applied for patents on aspects of the technique in August, two months after the births. Soon after we had begun to use starved cells, Keith saw that the technique might have other benefits apart from the convenience of being easy to stop the cells' development. Because of this we did not talk to anyone about the work, and so, after the lambs were born, we were able to submit a patent application. In the UK once an idea has been mentioned in public it cannot be patented; in the United States, by contrast, there is a year of grace during which patent applications can be submitted even after a public announcement. This later enabled the Roslin to draw in millions of dollars to set up a company and support the institute's research. For some reason that we do not understand, it seems that in the cell's quiescent state it is easier to change the functioning of the chromosomes: they are amenable to being reprogrammed because this represents a necessary step on the path of cell differentiation. The DNA carries the instructions for the cell, but other mechanisms regulate the way in which they function – so-called epigenetic mechanisms. These determine whether a particular cell will function for muscle or brain or gut. During nuclear transfer these cells have to be reset so that the chromosomes are able to support development of an embryo.

This result was so important to us and to other researchers that, after the birth of Megan and Morag, we were confident that one day it would be possible to clone an adult, and we began to look at just how we might do this. In science, the ability to plan this way is a tremendous advantage. That we were able to do so showed that our success rested on a deep understanding of the cell cycle obtained by Keith. The ability to use more differentiated cells for nuclear transfer gave us opportunities to carry out genetic surgery on the cultured cells before cloning, allowing us to create GM animals more efficiently. They also inspired other groups to investigate nuclear transfer. The next step – to be achieved within two years, I predicted at the time – was to 'design' a sheep by adding new genes to a cell before fusion, guaranteeing desirable characteristics such as an ability to grow quickly or to produce human proteins in animal milk that could be refined and used as drugs. All we would have to do was find one GM cell taken in a lab-grown line of millions and combine it with an empty unfertilised egg that could be harvested from slaughterhouse sheep.

Our experience with Megan and Morag also showed that reproductive cloning was risky. Of 244 attempts, they were the only survivors. We wondered whether we had been lucky or unlucky – whether the real efficiency was much worse or much better – and decided to repeat the experiments to find out. Both clones were later offered the chance to mate, though only Morag seemed to take up the offer and go on to have lambs. Alas, Morag succumbed to the same lung disease that afflicted Dolly, though Megan has lived on to the ripe old age of ten and counting. Davor Solter wrote an editorial to accompany the publication of the paper in *Nature* in which he stated that 'cloning mammals from adult cells will be considerably harder, but can no longer be considered impossible'. The media were not slow to pick up on the implications (except, notably and surprisingly, in America). As the *Daily Telegraph* reported in Britain, 'In theory, it is now feasible to clone human offspring without the participation of a man.' Lord Winston, one of the country's leading figures in reproductive science, was at pains to point this out: 'I don't think the human male is redundant. I think the human male is unnecessary.'

With the help of colleagues at neighbouring PPL Therapeutics, we cloned Dolly in 1996 and unequivocally realised Spemann's dream of decades earlier. The streamlined accounts that appeared in newspapers do little justice to how we restored to adult cells the full potency and potential of the embryo. Just dwell for a moment on what we were aiming to do. We were trying what, for most people, was unthinkable: to achieve a live birth without the help of a father. That meant carrying out complex surgery on an egg which is so small that you can only just see it with the naked eye. Whether whale, mouse, sheep or human, the eggs used for cloning mammals are approximately a tenth of a millimetre in diameter – around one two-hundredth of an inch. Now imagine trying to turn two cells – the egg and another cell – into one. Then we had to persuade the resulting reconstituted embryo to divide successfully. And we had somehow to reprogram the cell so that its DNA did not 'remember' ever having been an adult cell, but was sent back in cellular time to the embryonic stage.

We operated on sheep eggs using an astonishingly blunt instrument – a microscopic pipette with an inside diameter about the same size as the egg itself. We removed the egg's DNA, a molecule measuring 2.5 billionths of a metre across, and replaced it with the usual genetic complement of a cell, 2.3 metres or so of DNA, about eight picograms in all – eight trillionths of a gram. Ultimately, the effects of this surgery would have consequences at the molecular level: the egg is packed with thousands of complex structures, such as the mitochondria that provide it with chemical energy, and millions upon millions of molecules.

Astonishingly, we found a way to fool the egg, by a shock of electricity, into thinking that it was a developing embryo. Within each cell of a normal embryo, complex mechanisms normally ensure that genes are turned on and off at the right time in active regions of the nucleus, a membrane-bound region at the heart of the cell, to lay down tissues, organs and other structures. To succeed, we therefore had to find a way to persuade the egg to turn on the genes in the newly introduced DNA that are usually active in an embryo, while turning off those that work only in adult cells, and all without disrupting the careful choreography of genes that subsequently blink on and off during years of development. Put this way, the hurdles sound insuperable. Yet we succeeded, because of our finding that starving the donor cells was crucial if the genetic choreography within the reconstructed embryo was to have a chance of developing normally. We managed – albeit imperfectly – to reprogram an egg with adult DNA when we really had only the vaguest idea of the molecular details of the miracle of development. An 'impossible' lamb was born. This is how we made Dolly.

Chapter 4

Dollymania

Squash balls, infinite patience, a cramped, warm room, and udder cells from a long-dead sheep are just some of the ingredients of animal 6LL3. With the help of an egg, glass tubes little thicker than human hairs, and a flock of ewes, a big team at the Roslin derived every one of Dolly's cells, from her nose to her hooves, from a single udder cell. We still do not understand many molecular details of the cloning process, but despite these huge gaps in our knowledge the implications of this feat for science, medicine and society are huge.

One of the twists in the story of Dolly is that in all of the media hubbub that surrounded her birth and the acres of newsprint devoted to the possibilities of cloning humans, exactly how we cloned her and why we did it were examined in only the most cursory fashion. Few commentators seemed to realise that we did not set out to create Dolly at all, at least not as a primary objective, when we began in 1995. She was actually a detour in our quest to genetically alter farmyard animals to make new breeds that are healthier and more productive.

As we began to plan our next steps after the birth of Megan and Morag, several ideas were in the air. Could we use Keith Campbell's innovation of inducing quiescence to clone from even more differ-entiated cells – from foetuses or even from adults? There was also obvious commercial interest in the idea of adding genes. Our neighbours at the Roslin, PPL Therapeutics, wished to add genes to sheep cells while they were being cultured – perhaps even cells from embryos of the kind we had used to produce Megan and Morag. Significantly, as it would turn out, PPL also had to hand a line of cultured, *adult* sheep cells, something that my group lacked, for its studies of how to persuade sheep to produce drugs in their milk. We assembled at the Roslin in the autumn of 1995 to study the possi-bilities.

Alongside me in the group that managed the project were Keith, our long-haired cell cyclist, and Jim, the Roslin's stem cell expert. From PPL there were Alan Colman, their research director, who had done his doctorate with John Gurdon, the pioneer who had failed to clone adult frogs from adult cell nuclei. (Alan later con-fessed that he would have vetoed our work with adult cells if there had been embryo alternatives: when Dolly was born, Gurdon him-self said he was shocked.) Also from PPL was Alex Kind, who would go on to combine cloning with precise genetic alteration, and his wife, Angelika Schnieke, a Ph.D. student who at a critical stage in the sheep season came up with the good idea – backed with per-suasive arguments – of using in our experiments the cultured mammary cells that had been collected from a pregnant ewe.

Many others at the Roslin were involved in Dolly's creation, such as our cell surgeons. Karen Mycock, a twenty-eight-year-old, had already cloned cows for a company called Grenada Biosciences in Newcastle. Bill Ritchie, a neat man of military bearing, who likes to go 'Munro-bagging' – climbing the 284 Scottish hills that rise to over three thousand feet, a list first compiled by Sir Hugh Munro. There was also Bill's wife, Marjorie, who organised the surgery and operated on the sheep amid a steady supply of banter that kept the mood convivial and relaxed in the operating theatre.

Our multipliers of cells were Patricia Ferrier, Ray Ansell and Angela Scott. Mike Malcolm-Smith helped with surgery, and Douglas McGavin and Harry Bowran injected the sheep. John Bracken, a very experienced stockman with the quiet demeanour and that 'firm but gentle' manner that all animal handlers must cultivate, chose the name Dolly as soon as she was born. Standing in the pen, John said to Douglas McGavin, 'You know what we are going to have to call her, don't you? In view of the choice of cells for the experiment there is only one possible name: Dolly.'

The British government's Ministry of Agriculture, Fisheries and Food (MAFF) had agreed to fund our work on foetal and adult cells after we had succeeded in using the more immature cells. In the grant application, submitted two months after Megan and Morag were born, I had stated that the cost of an experiment to clone embryonic, foetal and adult cells would be an implausibly precise £484,519. A helpful civil servant played a key role at this point in Dolly's story. The project was funded only because of strong personal support from John Caygill, the plant scientist who was our project officer in MAFF. We got off to a good start in the autumn of 1995 and transferred nuclei from foetal cells and one group of embryonic cells. But PPL did not produce as many cultured embryonic cells as planned. So it came about that in early 1996 we had the time, the sheep and the money to do other things, and we decided to work on adult cells. Thus began the effort that would culminate in Dolly.

The Manipulators

The creation of Dolly entailed surgery on an egg much smaller than the tip of a sewing needle. To carry out this microscopic surgery, we needed people with steady hands, good coordination, persistence and patience. That ruled me out. Over the years, however, Roslin had trained a fair few 'manipulators' to do such fine work. Bill

Ritchie and Karen Mycock, who did the cell surgery in the Dolly project, have the unique skill of being able to concentrate for hours as they stare into a binocular eyepiece, yet remain relaxed enough not to make mistakes.

Some manipulators enjoyed listening to music as they worked on embryos. While pursuing his doctorate, Pedro Otaegui would listen to Leonard Cohen's downbeat songs of longing and despair. Bill and Karen would work side by side in a tiny room – a glorified cupboard – on the ground floor of the Roslin, one doing the cell surgery as the other fused cells to create the embryos. 'We used to sit and talk about a lot of rubbish really, or gossip, to pass the time,' said Karen. 'We could be in there for hours. It was like being in an isolation tank.' Since the work was done during the short winter days, 'sometimes you did not see daylight. You went into the lab during darkness and left during darkness.'

The 'micro' part of the micromanipulator rig that Bill used refers to how the cells that were to be combined by cloning were magnified with the help of lenses and optics. The microscope stood on a desk that was mounted on a heavy metal plate, which in turn sat upon squash balls to absorb the smallest vibrations. Without a way to soak up unwanted movements, a passing lawnmower outside the laboratory window, the slamming of a nearby door or even a racket from a radio would have made his cellular surgery shake, rattle and roll. The 'manipulator' part of the instrument's name refers to its two arms, each of which brandishes a fragile, hair-like pipette through which oil can be gently sucked or blown to pick up a cell or put it down. On one side is a pipette one-tenth of a millimetre in diameter, which is almost the same diameter as the egg it is used to grasp. On the other is a much smaller pipette, which can be as small as one-thousandth of a millimetre in diameter and is used to remove DNA and to move the donor cell about.

Bill moved the pipettes by twiddling two knobs (more modern versions have a joystick like those used with computer games, only more precise). To pick up the egg, you apply suction with a syringe. To set it down again, you release the pressure by pushing in the plunger of the syringe. Micromanipulation is as fiddly as it sounds, requiring patience and months of practice. A high degree of coordination is essential and the work is extremely repetitive – Bill said that it needed as much concentration as flying a helicopter. 'Three hours and you're knackered.'

The Egg Hunt

For cloning, you first need a reliable supply of eggs – hundreds of them, if possible – because nuclear transfer is very hit and miss. Although females carry a lifetime's supply, the eggs must be at the right stage of development. In a mother, whether sheep or human, eggs are ripened one by one in a cycle of hormones, gradually depleting her supply. After birth her ovaries contain many thousands of eggs, each wrapped in layers of cells, in a structure known as the follicle. During the weeks leading up to ovulation, the egg and its surrounding follicle are matured by changes in hormones. In the days immediately before ovulation the egg makes proteins, RNA and other molecules whose job will be to direct development just after fertilisation.

Today we can take a slaughterhouse sheep ovary and ripen the eggs that it contains in the laboratory. But in the early days of our research, from 1985 until 1995, we had to work much harder and obtain ripe eggs by carrying out surgery on a ewe at a precise moment. Because sheep give birth in the spring, when their lambs are most likely to survive, and because the gestation time is around five months, Dolly's story begins in midwinter, when our technicians and shepherds rounded up the ewes – a cold, wet, muddy, hard and smelly job. Each day they had just six hours of mean midwinter light to work by.

We used around sixty donor ewes to obtain the eggs, carefully orchestrating their hormonal cycles. Each ewe was first 'sponged': a sponge impregnated with the hormone progestogen would be put in the vagina to keep the ewes from ovulating. We then injected another hormone, gonadotrophin, twice a day for four days to make the ewes produce more eggs than the natural number of two or three. A final jab would make the animal ovulate at the appropriate time, within a six-hour window, so that we could be ready to remove the eggs.

In the Large Animal Unit at nearby Dryden Farm, two of our team would clean the sheep, shave them, disinfect their pink bellies and anaesthetise them. Unlike cattle, which have to be moved with a hoist, a sheep is light enough to be lifted onto the operating table without mechanical assistance. Another two members of the team would begin the effort to flush out the eggs, one operating and one assisting by passing instruments and so on. A fifth member would hunt for the eggs and pick them out of the sterile fluids that had been washed through the reproductive tract of the sheep, hopefully flushing out the eggs. I enjoyed the camaraderie of this team effort, which went on for two or three days a week over a period of several weeks. I did not enjoy the bureaucracy. A lot of paperwork was required to obtain funding from several different agencies, report on past and ongoing projects, and satisfy the Home Office that we had treated the sheep humanely.

Bill and Karen had to use eggs at just the right stage of development for nuclear transfer to have the best chance of success. When an egg ripens in a ewe's body, it must undergo meiosis, a strange form of cell division that carries out vital bookkeeping when it comes to the numbers of chromosomes that the eggs (and sperm) carry. We would collect the eggs when the ewe was in heat, when we knew that the eggs were part of the way through meiosis. (The details are more complex than many realise, and readers of a nervous disposition may want to jump to the next section.)

The vast majority of body cells in a sheep package their DNA into twenty-seven pairs of chromosomes (we have twenty-three pairs in our cells). They are called diploid cells, and they multiply by another process of cell division, mitosis. But for an egg and sperm to combine to get a new life going, they each have to bring half that number of chromosomes along to create the usual quota of chromosomes in an offspring. Sheep eggs and sperm start off with a normal number of chromosomes and then, by the process of meiosis, throw out half this number; they are thus left with twenty-seven chromosomes and become what are called haploid cells.

Meiosis can be thought of as a variation on the theme of normal cell division (mitosis). Two meiotic cell divisions of a diploid germ cell make four haploid sperm. But meiosis is different in eggs: half the chromosomes are ejected in what is called a polar body – a tiny cell that lacks enough cytoplasm to function. Thus, while the cell division in mitosis creates two viable daughter cells, meiosis in eggs creates only one. It gets more complicated, and scientists divide meiosis into several stages. There are two cell divisions called meiosis I and meiosis II, which cull chromosomes, each of which is divided into further phases in which different chromosome 'dances' take place within the egg – prophase, metaphase, anaphase and telophase.

We discovered that one stage of meiosis works best for cloning: the second metaphase, just before the egg is shed from the ovary. At this stage, rather than being uncoiled in the nucleus, as it is when its instructions are being read, the egg's DNA is conveniently packed in short chromosomes that are assembled on a cellular structure called a spindle, ready for the next phase of cell division, when some of chromosomes are shed into a second polar body after the sperm penetrates.

Once the precious eggs were collected at the Large Animal Unit, they had to be taken back to the Roslin Institute so that cloning could begin in earnest. They were placed in a standard little plastic test tubes with a cap, called an Eppendorf. Bill would pop the Eppendorf into his shirt pocket; Karen, lacking the pocket, would tuck the container of eggs under her bra strap to keep them warm. 'You did not want them to chill,' she said.

Breaking Eggs

To make Dolly required teamwork of military precision on those chilly winter days. In order to make the ewes shed several eggs and to control the timing, they were given a series of injections. Typically we would give the ewes the last injection on the Monday and recover the eggs on the Tuesday morning. That afternoon Bill and Karen would get to work on them. On the same day we also had to remove the fruits of the preceding week's work – cloned embryos – from sheep that had acted as temporary recipients to allow the embryos to develop. The interlinked cycles of work on the sheep in the Roslin somehow reminded me of the vine that winds itself around one of the pillars in the nearby Rosslyn Chapel. For Christians it was the tree of life; for the Norse it was the tree of knowledge, Yggdrasil, on which Odin sacrificed himself to gain the secrets of creation.

Bill remembers how Dolly was made during the last few weeks of the season and how it was touch and go whether she would be made at all: 'There was talk of doing it the next season.' Karen recalls that the day on which Dolly was created 'was quite a slow day' as there were not many eggs to work with. For this little act of creation the eggs had to be kept happy, which meant carefully controlling their environment. The cramped room in which the team worked was kept warm. The eggs, like the cells used to donate DNA to the reconstructed embryo, were kept in a specially prepared culture medium to mimic the chemical balance in the ewe's body. Typically, this medium contained salts, energy sources such as sugar, proteins and amino acids. We also added antibiotics to stop bacteria from setting up home in the manipulator and ruining everything. To keep the medium slightly alkaline, as it is in the body (at a pH of around 7.4, similar to that in a swimming pool), we used a chemical buffer. Bill would place this medium in a special chamber on a microscope slide. The chamber was sealed with a slick of mineral oil to prevent fluid from evaporating. The pipettes were passed through the oil to carry out the cellular surgery.

Once he had placed the eggs in the micromanipulator, nuclear transfer could begin in earnest. To enucleate the egg (remove its DNA), Bill had to break through its protective shell, the zona pellucida. This shell is crystal clear, as the name implies, though not brittle like a hen's egg. Nature designed the zona as a gateway that could be penetrated by one sperm, but no more. The zona has a coating that allows sperm to stick on, and penetration is achieved with a great deal of tail thrashing along with a cocktail of zona-dissolving enzymes in the sperm head.

The elasticity and toughness of the zona varies from one species to another, which makes a big difference to the nuclear transfer process. In the case of Dolly, we penetrated the zona using pipettes with sharp tips. Sometimes we used a stepwise jab by exploiting the piezoelectric effect, whereby certain crystals can be made to distort when a voltage is applied to create imperceptible movements, to move the pipettes. This method was developed in Hawaii by Ryuzo Yanagimachi's laboratory to pierce the zona pellucida of the mouse egg, which is unusually elastic and hard to penetrate, compared with that of a sheep egg; we tried it to see whether it might offer any advantage.

Once Bill was inside the zona, he had to find and remove the egg's DNA, which was coiled up in the chromosomes. At this stage of egg maturation the chromosomes were grouped near the surface of the egg. Even though the chromosomes could not themselves be seen, Bill could get some idea of where they lay because they were

packaged together on the spindle (the structure that helps to separate chromosomes during cell division). However, they were still far
from obvious. Sheep, like other farm animals, have dark grey eggs
brimming with fatty droplets that obscured Bill's view.

In this fatty fog, Bill had to play hunt the spindle. One way was
to cheat by adopting a cunning strategy developed in the mid-1980s
by Steen Willadsen: gently slice the eggs in half. Because the DNA
is in a small lump, it ends up in one half of the egg. Unsurprisingly,
this radical solution cuts the size of the reconstructed embryo and
its precious cytoplasm that helps the whole process of cloning
along, and thus reduces the chance of achieving a pregnancy.
Another way to play the spindle game emerged a few years later.
Yukio Tsunoda of Kinki University used special dyes that bind to
DNA and glow blue-green in ultraviolet light. If the eggs were
incubated with one of these dyes before being exposed to ultraviolet light, it was possible to see the chromosomes clearly, even
when they lurked in the cloudy eggs of farm animals. Once again,
however, there was a catch: ultraviolet light damages the eggs and
makes them less viable, in much the same way that it wrinkles and
degrades the skin of sun worshippers.

At that time, the Roslin's own Lawrence Smith came forward
with a way, which we still use today, to track down the chromosomes. He found that he could spot them by looking for small
clearings in the cloudy cytoplasm. Because the DNA is clumped in
the spindle, the surrounding fog of fat lifts a little in this one spot.
This method works in about 70 per cent of cases. To be sure he had
removed them, Bill used Tsunoda's DNA labelling technique and
then screened the contents of the pipette he used to suck out the
egg's chromosomes: if it glowed under ultraviolet light, then we had
been successful in removing the labelled DNA. This way, we did not
upset the egg's cellular machinery by cooking it with ultraviolet
light. (Jean-Paul Renard and his colleagues in France have developed yet another approach, in which they use extremely sensitive

cameras attached to their microscope, enabling them to dim the ultraviolet light to such a low level that it can do no damage.)

With experience and patience, and Lawrence's method, an experienced manipulator like Bill Ritchie could remove DNA from up to a hundred eggs over a two-hour period, so long as the eggs were of good quality. They had a smooth outline, a uniformly cloudy cytoplasm, and were free of any clingy cumulus cells – the cells that nurture eggs during their development. (This measure of quality is, of course, partly subjective and one of many examples of the 'witchcraft' of nuclear transfer.) Crucially, however, Bill did not remove all of the contents of the egg: left inside was a jelly – the cytoplasm – that contains all the important chemical machinery to reprogram any newly introduced DNA. Now we had to obtain that DNA so that we could reprogram the empty egg.

The Udder Donor

We are now halfway through the process of nuclear transfer. The enucleated egg could be thought of a CD player that had just had its CD removed. To get the egg to play a new genetic tune, Bill required the equivalent of another CD, a new set of chromosomes. In the case of Dolly they were obtained from a cell in the mammary gland of an adult ewe. That is why we named her after the famously endowed Dolly Parton. None of us at the time could think of a more impressive set of mammary glands. The singer's agent later said that she was honoured by the tribute, joking, 'There's no such thing as baa-aa-aad publicity.' Her name was a gift for the headline writers: 'Cloned Sheep's Name Brings Back Mammaries', ran one. Another, which accompanied Dolly's announcement in the scientific literature, was 'An Udder Way of Making Lambs'.

The mammary cells had originally been obtained during collaboration between PPL Therapeutics and Colin Wilde at the Hannah

Dairy Research Institute, near Ayr, Scotland, to study lactation. When the tissue was removed by surgery, the six-year-old Finn Dorset ewe was part-way through pregnancy and under general anaesthetic. Years later, and long after her death, her cells were conveniently at hand for our cloning experiments. That is why, when Dolly was born, there was no original donor animal to which we could compare her; in retrospect, this would have been a useful thing to be able to do. We came to grow weary of being asked about the fate of Dolly's identical sister/genetic mother.

Dolly's mammary tissue was being used in research to increase milk production. One idea was to study modified mammary cells grown in the lab and select the ones in which more protein was made – which was thought to be doable, then to clone the best animal from it – which seemed infeasible at the time. As it turned out, the impossible part of this project worked and the plausible part did not.

We grew the donor udder cells in a culture medium that contained blood serum from foetal calves, chosen for the simple reason that it worked. That way we ended up with a large number of cells that were identical to the donor's. We did not use all of them for nuclear transfer: some could be frozen and stored, almost indefinitely, in liquid nitrogen, which has a temperature of minus 196 degrees Celsius and, better still for cash-strapped researchers, is cheaper than beer. This way we could make clones years after the cells were taken, and long after the donor died. We could make another Dolly today from that same groups of cells, though she could never fill the giant hoof-steps left by her older identical sister. Nor would we be allowed to do such a thing without a good scientific reason.

A healthy donor cell boosts the cloning procedure's chance of success. For Dolly we used one simple health check: we counted the cell's chromosomes. This is one reason why we cultured cells for nuclear transfer. While they were growing in the laboratory we could from time to time tot up the chromosomes in a sample of one hundred cells. If we found that a high proportion of cells had abnormal numbers of chromosomes, we threw the lot away. Critical to the success of our attempt to clone Dolly was the preparation of these donor cells before nuclear transfer. That meant using the cells at the right point in their cycle.

During each cell cycle there are precise mechanisms, still not fully understood, that come into play to ensure that during each period of cell growth the genetic information in the cells is not only copied but copied just once, and with great fidelity, so that cell division can succeed. In this way the daughter cells contain exactly the same DNA as the parent cells. As mentioned above, Keith Campbell was one of the first to emphasise the importance during nuclear transfer of ensuring that the cell cycle of the donor cell is coordinated with the cycle of the recipient egg. In effect, these cellular clocks must tick on the same beat. Usually this coordination is achieved by using cells at the beginning of the cycle, but this is difficult to arrange with embryonic cells. The cells that Jim McWhir grew from the embryos had the advantage that they would stop at the beginning.

We starved the mammary cells of serum for five days, and they stopped growing and dividing to enter the state known as quiescence, or G0. Intriguingly, this state had already been linked with differentiation, the process in which an embryonic cell evolves to specialise as a skin cell, muscle cell or whatever. This, in effect, gives the cell a breathing space so that DNA can be reprogrammed to change the cell's fate. The G0 state can also be thought of as a form of hibernation, and it happens slowly and smoothly: whereas an electric motor, for example, simply stops at the flick of a switch, a starved cell shuts itself down in a gradual, orderly way. The cells no longer copy their genetic information, and become less active. Scientists think that it is because the organisation of nuclei in these quiescent cells is special in some way that they are more receptive to nuclear transfer than if they were at other stages of the cycle of cell growth and division.

The reason that G0 had been neglected in prior work was that cells from early embryos will not go into this stage of their own accord. Indeed, if you starve cells from early embryos, they do not hibernate – they die. Since G0 is a form of hibernation, it was much easier to hold more advanced cells at this stage than at other cellular checkpoints. (We could have used another, G1, but G1 gave us a window of opportunity lasting between half an hour and an hour. To use this checkpoint would have required having several teams working in a carefully planned way to coordinate all the elements of nuclear transfer.) We were ready to go.

Fusion

By now we have the nuclear donor and the enucleated egg – the CD player and CD, if you like. To slot the CD into the player and reconstruct the embryo for cloning, the DNA from one of these starved cells has to be introduced into our enucleated egg. One detail that never fails to catch the eye of journalists is that, to kick-start development, we used a shock of electricity. Given how this happened in the winter months, there was the inevitable echo of the moment of Frankenstein's creation. ('It was on a dreary night of November that I beheld the accomplishment of my toils. With an anxiety that almost amounted to agony, I collected the instruments of life around me, that I might infuse a spark of being into the life-less thing that lay at my feet.')

First, Bill Ritchie would inspect the mammary cells and select a diminutive specimen, which was more likely to have been starved and thus be near the start of the cell cycle. He checked that the cell had a normal surface, without a surrounding halo – another sub-jective sign of good health. Then he took the small pipette that he used for enucleation to pick up and pass the donor cell through the zona pellucida, usually through the same opening used for enucleation. The donor cell, like most cells in the body, is much smaller than the egg and could pass into the narrow pipette. Bill deposited the donor cell next to the enucleated egg inside the zona, which was enclosed in its own inner membrane (like the contents of a hen's egg under the shell). Because he had ten to twenty eggs in each chamber of the micromanipulator, he repeated this step ten to twenty times before taking them, en masse, to a second microscope to fuse their contents together, so that the egg could now 'play the new CD'.

Fusion was then carried out by Karen Mycock. It took place in a special glass dish, a few centimetres across and a centimetre deep. This amazing feat of cellular intermingling was achieved on the bottom of the dish, where two wires were glued one-fifth of a millimetre apart. We had bathed the wires in a special salt-and-sugar solution, like the reconstructed eggs, so they made a good electrical contact. First Karen would prime the cells with a burst of alternating current, creating a different charge on the two cells, which drew them closer together until they kissed. After the a.c. came the d.c. She passed pulses of electric current through the two cells to initiate the process in which they would merge. This is more difficult than it sounds. Other teams had shown, through trial and error, that for fusion to succeed the electric field must be perpendicular to a line between the centres of the cells at the point where the cell membranes touch. When oriented correctly, the d.c. electric field drills holes in the cell membranes, enabling the contents of the cells to mingle.

This fooled the reconstructed egg into thinking that it had been invaded by a sperm, a process called activation that is linked to the release of calcium. In effect, Dolly – and you and I too – were conceived in a flash of calcium. After fertilisation one can detect pulses of high calcium concentrations every few minutes that continue for several hours. The calcium triggers a chemical chain reaction throughout the fertilised egg that allows an embryo to form, and also denies access to any other sperm by changing the properties of the zona, making it tougher, so that it can – in the body – resist the advances of any more sperm.

The calcium explosion throughout the egg's spongy interior also helps the successful sperm to overcome a logistical problem. The egg is thousands of times bigger than a sperm, but the whole egg must undergo activation at fertilisation. In the absence of the sperm, there are various ways to do this, which all mimic the effects of the calcium pulses in one way or another. To create Dolly, Karen used an electric shock to trigger the release of that all-important calcium locked up in cellular stores and initiate cell division.

After activation, Karen put the reconstructed egg into the equivalent of intensive care: a culture medium where recovery could take place. As the contents of the enucleated egg and donor cell repaired the holes, the two cells merged into one. As Karen watched, she could see fusion gradually take place. Even after half an hour the smaller donor cell could still be seen nestling against the egg. After an hour or so the donor cell – and its DNA – had been swallowed up by the egg and its own outline was no longer visible, rather as two beads of mercury become one. Dolly was one of the 70 per cent of cases where the cells fused successfully. (There is now an alternative technique for introducing the donor nucleus: the donor cell is sucked against the tip of a pipette that is smaller in diameter until the cell bursts. The contents are drawn into the pipette and then injected directly into the egg.) Karen and Bill would joke that it was they who were really Dolly's mum and dad. 'I was the dad, and he was the mum,' she says.

We expected that within a few days of reconstruction the single cell of an embryo should divide several times to produce first two, and then four, eight and sixteen cells. Dolly was one of the 50 per cent of embryos that seemed to be developing normally when all went well (sometimes not one embryo would develop). At the six-teen-cell stage, which is reached after four days, the cells are independent but still held together by the zona to form a ball. The membranes of the cells on the outside of the ball stick to one another and begin to control the access of molecules to the cells in the middle. At this stage an embryo is known as a morula, and we begin to see the start of what really sets mammals apart from bac-teria – the ability to divide the task of staying alive among a fully coordinated and cooperating community of cells.

Fluid begins to accumulate inside the morula to create a single fluid-filled cavity, at which time the embryo is a hollow ball of cells known as a blastocyst. There are between fifty and two hundred cells in the embryo at this stage, depending on their quality, and some-times in preliminary experiments we stain the embryos to count the cells and make the first comparisons between different treatments. This is the stage at which an embryo can implant itself into the wall of the womb in some species, such as human and mouse. However, the ability to grow reconstructed embryos to this stage in the lab is by no means a guarantee that an offspring will result. In their exper-iments on mice, Teru Wakayama and his colleagues at the University of Hawaii found that varying the time between nuclear transfer and activation made little difference to the percentage of embryos that grew to the blastocyst stage, but it was only with those for which there had been a certain delay that they managed to get cloned mouse pups. As ever, the details of how to succeed vary from species to species.

Planting Embryos

We were still groping for those important details when we created Dolly. We faced a major problem in growing reconstructed embryos up to the blastocyst stage, at which point they became mature enough to implant in the womb: there was no really effective method for keeping cloned sheep embryos alive for sufficiently long in the laboratory. The medium that nurtured the embryos of the most studied species, the mouse, did not work for sheep. The hard work to find the best recipe had not been done (teams in the antipodes led by Robin Tervit, Jeremy Thompson and Simon Walker have now solved this problem). If we tried to implant embryos before they got to be blastocysts, they were destroyed because the process of enucleation and nuclear transfer left them with a hole in their protective shell, the zona pellucida. By the blastocyst stage, the cells of the embryo has changed in some unknown way so that the embryos were not destroyed.

As ever, we turned to nature, exploiting another clever technique pioneered by that creative Dane, Steen Willadsen. For this, we needed a second set of surrogate mothers. To protect the reconstructed embryos from destruction, we encased them in a shell of agar. We tied off the oviduct (the equivalent of the fallopian tube in a woman) of another sheep and placed the encased embryos there, where they could develop for six days, to the blastocyst stage. This part of the operation would tend to fall on the Wednesday of a routine week.

The following Tuesday, when the embryo was about to outgrow its temporary shell, Bill Ritchie had to recover the embryos and remove their coat of agar – one of the most difficult parts of the operation. Once he had stripped it of its protective shell, Bill could reimplant the blastocyst in a second ewe. This ewe also had to be receptive to the embryo: half a century ago, it was first shown that if you moved an embryo from one animal to another, the two animals had to be at the same stage of their reproductive cycle for the embryo to survive. There is a chemical dialogue between blastocyst and mother, and this must not be interrupted. With the sheep under anaesthetic, Bill's wife, Marjorie, made an incision just in front of the udder to expose the ewe's ovaries. To make sure that the ewe was at the right point in her cycle, she would check whether the ewe had ovulated from both ovaries or only one. We tried to match the number of days for which the cloned embryo had been developing in the laboratory with the number of days that had passed since the recipient ewe was last in heat, then knocked off a day or two to allow for the fact that reconstructed embryos did not manage to grow as vigorously as the real thing (still more evidence that cloning has deleterious effects). In cloning jargon, egg and womb must be 'synchronous'. Work in the 1960s by Bob Moor and Tim Rowson at Cambridge had showed that a blastocyst has the best chance of developing to term if it is transferred into a recipient ewe that has been in heat six or seven days previously, as if it had carried that blastocyst for that same length of time. Then they are synchronous.

Embryos tend to develop more successfully in the part of the uterus that is next to the ovary from which an ovulation has taken place, perhaps because locally acting hormones help to pave the way for pregnancy. After three or four days of being propelled down the reproductive tract by tiny waving fronds (cilia), the embryo would normally reach the uterine horn. Because the embryo is then less than six days old, the spot where it had the best chance of survival was at the tip of the horn. First we passed a needle through the wall of the uterus. Through the same opening we inserted a glass pipette carrying the embryo. The embryo, inside a small volume of medium, was injected into the sweet spot. If all went well, at about fifteen days the transferred embryo would begin to attach itself to the wall of the ewe's womb, becoming fully implanted by the fifth week. All we could do then was wait and watch the ewes.

Waiting for Dolly

My colleague John Bracken, a soft-spoken man who has witnessed the birth of a dozen or so clones, would monitor the pregnancies

with Douglas McGavin. Both had spent a lifetime looking after animals, particularly sheep, and knew and understood every detail of
their care. For the Dolly project, limitations of the ultrasound kit of
the day meant that they could scan the sheep only two months into
the pregnancy. Instead, they had to use old-fashioned methods to
find out initially whether all the surgery and toil in the laboratory
had paid off. What they did was tempt the surrogate mothers with
'teaser rams' sixteen or seventeen days after the last time the ewe had
been in heat: if the embryos had failed to develop, the animal would
have returned to heat. After a couple of months, John and Douglas
could place an ultrasound probe against the belly of the surrogate
mothers, the Blackface ewes; it was the very same equipment that
was used for monitoring pregnancy in women. By this time the placenta had grown, and the uterus of the surrogate had filled with
fluid, which could later be seen as dark patches on the screen of the
scan.

In the first tranche of the experiments, we used cells from a nine-
day-old embryo (the same type we used to produce Megan and
Morag), differentiated skin cells from a twenty-six-day-old foetus
and, of course, those frozen udder cells. The first ultrasound scans, at
around eight weeks, showed that only one-third of the ewes had
foetuses. Much to our surprise and delight, however, one ewe was
pregnant with a clone from an adult cell. Ultrasound revealed an
apparently normal foetus. Karen Mycock recorded the scan on a
video with maternal pride and felt a thrill of anticipation that a
lamb could result from this supposedly impossible experiment.

That day, 20 March 1996, was when we first began to think seri-
ously about clones from adult cells. Keith Campbell and I became
nervous, particularly as the pregnancies continued to fail. Except,
that is, the one that should not have been: Dolly. We shared a mixture of excitement and concern because we knew that a milestone
would be passed if the lamb survived. We knew from previous experience that many of the foetuses die, sometimes late in pregnancy or

even at birth. As we fretted over Dolly the foetus, the Princeton professor Lee Silver was writing a popular book on reproductive technology in which he was patiently explaining why cloning from adults was a biological impossibility.

That May, around 110 days into pregnancy, four foetuses were found to have perished, all clones of embryos. Of the 29 embryos that we had created from 277 udder cells, one foetus continued to thrive in her Scottish Blackface surrogate mother. We constantly monitored her well-being. Every time John Bracken used ultrasound, it always took a few seconds to get the whole image in view. You could usually see the head, legs and ribs at first. There would also be some movement. Most important of all was the moment when we could see a heart beating. Then we all felt a great sense of relief and satisfaction.

To ensure that our nails were bitten to the quick, Dolly's gestation, like that of other clones, took longer than normal. In most British breeds of sheep the average duration of a pregnancy is 147 days. Experience now tells us that pregnancy sometimes drags on up to 155 days for clones. They usually die if the pregnancy goes beyond day 153. The labour of a ewe carrying a clone is also slow and sluggish, as we now know also to be the case in other species. Just before the onset of labour, a ewe having a normal pregnancy will often leave the rest of the flock and make a 'nest' by pulling grass or straw together to form a hollow in a quiet corner of a field, an ancient reflex to create a defence against predators. For reasons we don't understand, nesting was less likely to be triggered by a cloned embryo. The wait continued.

Dolly's Birth: More Bleat Than Bang

We did everything we could to keep the ewes and the clones safe, and placed the animals under twenty-four-hour observation. We wanted to leave nature to take its course, but if the pregnancy went beyond 153 days we would induce birth with an injection of the same hormones that the foetus normally releases to signal that it is ready for the outside world. The lambing pen at Roslin has seen great excitement and satisfaction, but more often than not with clones it has also witnessed death and deformity. But that was not the case on 5 July 1996. A healthy lamb was born that day, one that would become the talking point of paupers and presidents.

Dolly entered the world head and forelegs first in a shed on the Roslin farm late in the afternoon. Her arrival was a muted affair. She weighed 6.5 kilograms, surprisingly large for a normal lamb but not for a clone; we now know that animals cloned by nuclear transfer are often oversize. In attendance were a local vet and a few staff members from the Roslin. John Bracken was in charge. Even though I am often called the father of Dolly the sheep, I was an old-fashioned father and was not present at her birth. Her expected arrival was the cause of a great deal of excitement among the team, and I gave the instruction that only those who had to be present should be there. At the time of the birth I was digging in the vegetable garden in the grounds of the institute. One thing her mother did not need was stress brought on by too many people being present.

Bill Ritchie and Karen Mycock were thrilled with the idea that their cell manipulations had paid off. Bill was one of the handful of people there at the birth. 'She was one of the few animals I saw being born,' he said, and recalled how he did 'a bit of jumping up and down' when Dolly appeared. But, having made his first clone three years earlier, and given that Megan and Morag were probably as significant, Bill admits that it came as an anticlimax. Karen was at a friend's Highland wedding in Plockton that momentous Friday. Angela Scott had promised to keep her informed, since they knew that the birth was imminent. When Karen returned to her room at the Craigendarroch Hotel to change her hat, she found a fax from

Angela that read, 'She's been born and she has a white face and furry legs.' 'Heaven only knows what the receptionist thought as she slid the fax under our room door,' Karen later recalled.

Angela's message meant that the lamb was different from her Blackface surrogate mother. They could rule out any thoughts of a mix-up – they had a clone of an adult cell. Although Dolly had no name at that point, Karen knew that the lamb would make headlines: 'I tripped the light fantastic. I thought hurrah and was absolutely ecstatic.' At the ceilidh in the village hall that night, Karen bought drinks for everyone. 'I'm a dad,' she shouted. 'I think the locals will remember the "mad scientist" who claimed to be a sheep's father and then sunk more than one tequila slammer and proceeded to dance the night away.' Months later, Karen sent the bride and groom a copy of the Dolly paper so that they could see what the fuss had been about.

Keith, who had been so heavily involved in her creation, was not present for Dolly's birth either. He was with Ange, his partner, on a camping holiday in Devon that had been booked months earlier. After he rang in, as he did every day, and got the news, he cracked open a bottle of wine with her in a field to celebrate.

I did not hand out cigars or go for a celebratory drink in the local pub. No one took photographs. Nor could my wife remember me coming home and performing cartwheels of joy. Roslin was not painted red. Dolly's creation had been so long coming, so protracted, so difficult and such hard work that we were too fatigued to cry 'Eureka!' I had bought a bottle of champagne for the event, but it seemed wrong to uncork it without Keith. In the end, Alan Colman of PPL treated the entire team to a night out later in the summer when everyone was around.

There were other reasons not to shout about Dolly from the rooftops. Because we intended to submit the work for publication in *Nature*, we had to keep the news of her birth secret for many months. The journal has a policy of rejecting papers if the results are announced before publication. The manuscript was submitted on 22 November and, after minor amendments, accepted in January, ready for publication towards the end of February.

Brouhaha

We could see the media storm coming but were unprepared for its severity. We arranged for a public relations company, De Facto in Basingstoke, to prepare a press release and other material. I practised press interviews with Ron James, the biochemist turned entrepreneur who ran PPL, an immaculately tailored man with piercing blue eyes, white hair and a white beard. We tried everything, from the aggressive jab of the microphone up the nose to the traps set in lengthy conversations with writers in search of a new angle ('So you really cloned Dolly because you wanted to clone yourself, then?').

As part of its normal pre-publication process, *Nature* told reporters on Friday 21 February 1997 that the paper was to be published in the Thursday, 27 February issue, triggering a trickle of advance calls. This practice of providing embargoed information allows journalists to investigate the subject, carry out in-depth interviews and be well informed before publication. By this time, the *Observer* – published on a Sunday – was already on to the story and ran it on the 23rd, helped in part by a television producer whom we had told all about Dolly because his film was to be broadcast after her birth. The front-page story, by Robin McKie, contained these words: 'it is the prospect of cloning people, creating armies of dictators, that will attract most attention'.

I have no doubt that his hunger for a scoop made it more difficult for us to announce the result. We were only part of the way through a careful process of informing those people who might be asked to comment about what we had achieved: other biologists, ethicists, clerics – the pundits to whom the media turn for guidance and comment. Doubtless they, like us, were now being inundated with phone calls requesting views on something they knew nothing about until that Sunday morning. The premature publicity also caused difficulties within the Roslin Institute. We had planned to inform the staff members before the end of the embargo, but they heard what had been going on in their own institute from TV, radio and the newspapers. One of our first jobs on the Monday morning was to address the staff and reassure them that we had no intention of cloning people, but had many other important applications in mind. We also apologised for the news leak and told them what had been planned.

We found, despite all our hard work, that nothing could have prepared us for the thousands of telephone calls and the scores of interviews that followed. There were pronouncements from the great and the good. There was much angst. But, reassuringly, there was little outright hostility or opprobrium. As soon as it was clear that the story was breaking early, I went to the institute with Harry Griffin, the Roslin's bearded, affable and much respected assistant director, to take calls from the media. Ron James, who was in the South of England for the weekend, went to the Basingstoke offices of De Facto.

We could see the journalists of the planet wake up as the Earth turned on its axis. Reporters rang first from Europe, then from the East Coast of America, then from the West Coast. Before the media hordes in Australia and New Zealand had their chance, we had gone home, wrung out and exhausted. Hundreds of e-mails began to arrive. The next day the Roslin was besieged by British and American TV crews. Satellite vans in the car park beamed live interviews to morning news programmes on the other side of the Atlantic. Invitations poured in for conferences, lecture tours, books and films, and even for Dolly to make a personal appearance on an American talk show.

The press conference that we had planned for Wednesday was held at the Roslin on Tuesday morning, attracting a crowd even though the story was already 'old', having broken two days earlier. The story had legs, as reporters say – four legs, in this case. With four film crews and twenty photographers, Dolly coped well, as if she had been doing media work all her short life. I would be photographed with her many times over the years. Some city dwellers seemed not to understand that you cannot tell a sheep to stand in a specific place where their spotlights are focused, so we began to use food to encourage her to come to a particular position. She was a sociable animal, even by the standards of sheep.

We were already beginning to tire of the incessant questions about cloning people. They were repeated a thousand different ways, and many times over, even at the same press conference, but our answers were the same: we thought it was unsafe, we did not approve, we weren't interested. We became equally weary of cloning jokes: there was much play on William Blake's quote, 'Little lamb, who made thee?' Why, a significant udder, of course.

The next day, photographers from *Time*, *Newsweek* and *Scientific American* arrived. In a journalistic display of shock and awe, a trio of reporters from the *New York Times* flew in from Moscow, Germany and New York. I was surprised to find that at the age of fifty-two I had become a celebrity. The *New York Times* described me thus in a section called 'Man in the News': 'His hobby is walking in the mountains of Scotland. He relaxes with "a good single-malt Scottish whisky", and enjoys the quiet of his village near Edinburgh that is "so small you wouldn't be able to find it in an atlas".'

All the time, more TV crews were turning up, usually unannounced. A few weeks later, Pepsi-Cola offered me and my wife an expenses-paid trip in a private jet to a party with other luminaries at the Hôtel de Paris in Monte Carlo. We accepted, and were joined by Harry Griffin and his wife. I was also asked to appear in an advertisement for a European mineral water; I declined. A photocopier company came out with a cheeky advertisement that declared, 'Big deal. We've been making perfect copies for years.'

Our secretaries, Jaki Young for me and Frances Fame for Harry, bore the brunt of the telephonic avalanche in the days after Dolly's birth. Still besieged by invitations and queries and radio stations across the planet, we were forced to draft in additional spokespeople. The Vatican and other religious groups voiced their fears about cloning humans. President Bill Clinton asked his bioethics commission, led by Harold Shapiro, the president of Princeton University, to report within ninety days on the implications of our research. Dolly, said the president, 'raises serious ethical questions, particularly with respect to the possible use of this technology to clone human embryos', and he called for a worldwide moratorium.

His statement carried strong echoes of the debate that greeted the birth of genetic engineering when, in 1975 at Asilomar, California, scientists in effect called for a five-year moratorium on the new technology until biological understanding and regulation caught up. They were concerned, among other things, that new techniques of gene splicing and cross-species genetic recombination might convert harmless microbes into virulent pathogens. Paul Berg, the Nobel laureate who proposed the moratorium, summed it up as follows: 'When you honestly don't know the implications of your research, what are you supposed to do? Just keep on doing it? No, you have to stop and ask yourself what the hazards really are.'

Even today, I am still trying to work out the hazards of cloning. But on one point I have remained firm. Within days of Dolly becoming global news, I came to dread the pleas from bereaved families asking whether we could clone their lost loved ones. I have two daughters and a son of my own, and I know that every parent's nightmare is to lose a child, and that I would consider anything that might possibly bring them back if it happened to me. But I had and have no power to help. Cloning does not solve the problem of bereavement. Even if you did clone a twin of the lost child, it would be a different person, and that would be cruel to both child and parents. For that reason, and for many others, I welcomed any move to ban the cloning of babies.

The Dolly Doubters

The corollary of the surprise that greeted the announcement of Dolly's birth was the scepticism and the doubts. These were voiced by some eminent figures. I cannot help feeling that they were fuelled in part by the realisation that Dolly was cloned by an obscure Scottish laboratory, rather than an American powerhouse of animal genetics (as would happen again, when the first human embryos were cloned in South Korea). Then again, I would probably feel the same if a major advance in my field came from beyond my peer group.

The donor ewe was pregnant at the time her mammary tissue was taken, and that provoked a question from Norton Zinder, a distinguished microbiologist at Rockefeller University, New York. With Vittorio Sgaramella of the University of Calabria, in Italy, he asked whether it was possible that Dolly could have been the product of a foetal cell that had been circulating in the donor ewe's body and not an adult mammary cell at all. The *New York Times* tasted blood and scandal. Zinder had indeed made a good point, one we had not even thought about. However, PPL had tested circulating foetal cells to see whether a pregnancy was going to produce a truly transgenic animal. It had failed to find any cells, suggesting that Zinder's scenario was unlikely. And the Finn Dorset cells, from the Hannah Dairy Research Institute, were the only preparations from that breed in all the laboratories involved in creating Dolly. Nonetheless, never say never.

We were lucky: someone from the Hannah rang up Alan Colman

to say that they still had some of the mammary tissue used to make Dolly. My colleagues busied themselves with DNA fingerprinting tests to confirm that Dolly was the sheep we said she was. Our critics, in turn, demanded to know why these tests had not been called for before publication in a peer-reviewed scientific journal of the stature of *Nature*. Scientists in my field of embryology were less rigorous than the molecular biologists who dominated the doubters, sniffed an article in the *New York Times*.

We upheld Dolly's provenance. Esther Signer of the University of Leicester provided independent DNA profiles confirming that Dolly was a clone of the six-year-old donor ewe. Signer, who worked with the inventor of DNA fingerprinting, Sir Alec Jeffreys, said that the results proved 'beyond reasonable doubt that Dolly is indeed derived from a cell of the mammary tissue taken from the adult donor ewe'.

There was another problem. At first Dolly was a 'clone alone'. The *New York Times* was troubled that no one had managed to repeat the feat by the first anniversary of the announcement. In fact, such a short timescale would have been unrealistic if you consider how long it takes to raise funds, buy the sheep and then carry out the experiment. By the first anniversary there simply had not been enough time to reproduce the feat. Even so, in March 1998 Jean-Paul Renard and his team near Paris made it clear that they had cloned a calf from differentiated cells, and that summer Ryuzo Yanagimachi of the University of Hawaii, and his postdoctoral colleague Teruhiko Wakayama, reported having cloned over fifty mice by nuclear transfer, including clones of clones.

Since then, research groups around the world have reported the cloning of various species, including cattle, sheep, mice, goats, dogs, cats, horses, pigs, mules, white-tailed deer, carp and fruit flies. Lacking as we do a fundamental understanding of how to clone, a lot of trial and error was required to customise the Dolly method to suit each species. Successful teams found the feat difficult in some species, notably the rhesus monkey and the dog. But probably any mammal can be cloned, given enough research.

Dolly's Little Secret

Dolly, like so many other 'clones', was not a true clone. Like most other female clones, she was made with an egg from one female and DNA from another. To have created a true clone, she would have to have been made by means of an egg and DNA from the same long-dead Finn Dorset sheep. Eggs contribute more to the development of the offspring than just the genetic information locked up in their chromosomes. Other genetic information is carried in the power-houses of the cells, the mitochondria we encountered earlier, which lie in the cytoplasm of the egg.

Some 3 per cent of the DNA in a cell can be found in the mito-chondria. While there is much less information in the mitochondria than in the chromosomes, it is crucial for normal development. Mutations in mitochondrial DNA accumulate with age more quickly than in nuclear DNA – for instance as a result of the action of highly reactive chemical agents called free radicals that are natu-rally produced in the body. If cloning affected mitochondrial DNA, it could be significant, as was shown in a study by Nils-Göran Larsson and colleagues at the Karolinska Institute, Stockholm. He bred mice with a defective version of DNA polymerase, an enzyme responsible for maintaining mitochondrial DNA, so that they suf-fered more mutations. These 'mitochondrial mutator' mice lived for only ten to twelve months, compared with up to three years for normal mice, and they aged prematurely.

Abnormalities in function of mitochondria can also lead to dis-ease. Richard Lifton of the Yale University School of Medicine found a genetic defect in mitochondria linked with hypertension, high cholesterol, high triglycerides, low magnesium, diabetes, insulin resistance and obesity (the defect affected transfer RNAs, critical carrier molecules that ferry amino acids around during the constructions of proteins, contributing to a broad range of cellular malfunctions). Given their role in powering the body, the first manifestations of having aged mitochondria would be in organs that have a high demand for energy, and are thus more dependent on mitochondria, such as the heart and brain.

Although we have yet to discover all the details of how the egg's mitochondria affect the clone, there is plenty of indirect evidence that they are important. Take a number of eggs from the same ewe and, after adding sperm and reimplanting, you will find that some produce healthy lambs, some do not develop at all, and others do develop for a little while but ultimately the embryo dies. Sometimes these differences in the viability of eggs will be down to differences between their mitochondria. But they are not the whole story. Apart from the influence of the mitochondria, events in the earliest stages of embryo development are also influenced by proteins and other molecules in the egg that were made before the egg even left the ovary. Very few of these active compounds have been identified, nor is their function understood.

Kamitakafuku

'Kamitakafuku' is the Japanese for 'God-given high happiness', as well as the name of a Black bull who was highly prized for his marbled, fatty 'wagyu' meat. By the time he was seventeen, his semen had been used to inseminate more than 350,000 cows, and this prodigious father had become a star in Japan, where bulls of his calibre are worth a fortune. In a conventional breeding programme it may take seven years to select a genetically elite stud bull like Kamitakafuku from hundreds of his peers, because of the lengthy process of testing for desirable traits. Rather than purchase the semen of genetically elite bulls to inseminate hundreds of cows in the hope that calves go on to develop these traits – despite the dilution of these elite genes with those of another animal – the beef industry hoped instead that cloning would create valuable copies of elite bulls without all the testing and selecting of progeny. In the same way, cloning could also be used to speed up the development of breeds suited to harsher conditions in developing countries.

Kamitakafuku died in 2001. But his genes lived on in more than the hundreds of thousands of offspring he had sired. He was the subject of an early mammalian cloning experiment that began in 1997, the year after Dolly was born. Xiangzhong (Jerry) Yang, of the University of Connecticut at Storrs, and Chikara Kubota, of Japan's Kagoshima Prefectural Cattle Breeding Institute, removed ear cells from the bull for nuclear transfer and in 1998 reported the births of six offspring. Four calves survived, including the first male clones of a farm animal. This initial set of cloned offspring – Saburo, Taro, Jiro and Shiro – developed normally into adults. Saburo and Shiro fathered offspring through artificial insemination, as well as in the old-fashioned way. Then, using ear skin cells from these clones, the team once again performed nuclear transfer. Nineteen embryos were transferred to nineteen recipients, which led to two live births of second-generation cloned bulls, on 23 January and 6 March 2000. Although one of the two calves died of anaemia and infection shortly after birth, the second lived on in apparent good health. The survivor was given the name Sho-zaburo and the English name Dr Lederman, after Jerry Yang's doctor, who routinely joked that he would like to be cloned.

The story took another twist in 2005 when Saburo and Shiro were slaughtered and their meat tested. Understandably, Jerry Yang was reluctant to do this, given all the effort his team had put into their birth in the first place. But because of the abnormalities caused by cloning, some scientists had raised theoretical concerns about the safety of eating clones. Yang's pilot study was the first to examine specific proteins and nutrients in the meat from cloned animals, and marked the start of efforts to fill an important gap in the research literature and achieve regulatory approval for clone-derived food. With Cindy Tian, Chikara Kubota and other colleagues, Yang had also cloned a particularly productive Holstein dairy cow (called Aspen), then compared the meat and milk from the clones with those from animals of similar age, genetics and breed created through natural reproduction.

'We conducted extensive comparisons of the composition of milk and meat from somatic [adult cell] cloned animals to those from naturally reproduced comparator animals,' he said. 'We found no significant differences.' Although many more studies would be necessary to convince regulators, the data provided the first scientifically obtained information to address public concerns about the safety of meat and milk from somatic animal clones.

This kind of work is also encouraging when it raises the hope of scientists, like me, that breeding can be made more precise if cloning is combined with genetic modification. Inserting genes offers a way to enhance breeds, adapt them to different environments, or alter them so they can grow 'humanised' rejection-resistant organs for transplant. They could even be designed to manufacture precious drugs in their milk. One could clone from cultured cells that had been transformed genetically to make a drug as they were grown. Beyond Dolly, an exciting landscape of new scientific opportunities beckoned.

Chapter 5

Farmyard Clones

Our first attempt to blend the techniques of cloning and genetic modification rested in part on the solution to a bloody mystery published half a century earlier, when a paper appeared in the 1952 Christmas edition of the *British Medical Journal.* Working at Oxford University, Rosemary Biggs, Gwyn Macfarlane and their colleagues had found a new form of haemophilia, the inherited disorder that causes recurrent, uncontrolled bleeding.

An important clue that there was more than one form of the clotting disorder emerged when it was discovered that the blood of some sufferers could clot the blood from a patient who had the more common form, haemophilia A. Later it was shown that this variation of haemophilia was caused by the lack of a different clotting protein, one called factor IX. The Oxford team decided to name this new form of haemophilia Christmas disease after their very first patient, a five-year-old boy called Stephen Christmas, who had by then moved to Toronto in Canada. The journal rushed to get the paper into print at the appropriate time.

There was a precedent for naming disorders this way, set by the British surgeon Sir Jonathan Hutchinson. Indeed, Stephen's father, an actor, had himself played with the Yuletide theme. He named his older son Robin, and when he remarried he named the daughter of that union Holly. But, given that seasonal papers in the Christmas edition of the *BMJ* were traditionally flip and light-hearted, some doctors were unhappy. They had been expecting a meditation on overeating, drinking to excess or some festive frivolity. Some said that to name such a serious condition after a Christian festival was tasteless. Others dismissed the paper on Christmas disease as a student prank. One professor wrote to the journal to complain: 'Some parents live to regret the names they have thoughtlessly bestowed upon their innocent children. Would it not be possible . . . in your capacity as registrar of its birth, to substitute some less ridiculous name?'

The Oxford team stoutly defended the name but did assure the critics that if they found the protein used in the body to make factor IX, they would resist the urge to call it Christmas Eve factor. The only alternative name that the Oxford team offered was the dry, functional 'hereditary hypocoprothrombinaemia'. Other candidates were 'parahaemophilia' and 'pseudohaemophilia'. Unsurprisingly, 'Christmas disease' stuck, though today we tend to use an alternative, 'haemophilia B' (suggested by a distinguished scientist while all the arguments were going on). Unfortunately for Stephen Christmas, however, there was only one way to get hold of the clotting agent that he lacked: from human blood.

Half a century later we found a different way to make factor IX, in the milk of a sheep called Polly. Her birth was the first demonstration that cloning would also make it easier than ever before to carry out the genetic alteration of animals. In many ways her birth thrilled me as much as that of Dolly. Once again, our work was controversial: there are many who believe that Polly was another example of scientists 'meddling with nature'. Others found the idea of putting human genes into animals disconcerting.

But I felt, as I am sure did all my colleagues at the Roslin and at

PPL (where Keith Campbell was working by the time the details of the feat were published), that our attempt to turn Polly into a woolly drug factory was a natural extension of previous efforts by many people to make useful products from biological processes – biotechnology – that dates back thousands of years, to the first fermenting of wine and the origins of agriculture. The aims of the early farmers were little different in spirit from what we wanted to do. I would also like to think that Stephen Christmas would have approved of what we were attempting.

When we unveiled Polly along with four other cloned lambs, we felt that we had passed an important milestone in the long effort to adapt animals for life on the farm, offering new possibilities in the process. Indeed, we hoped that Polly's arrival would mark one of the biggest advances in breeding since animals were first domesticated, many millennia ago, because the techniques used to create her made it possible to alter animal DNA in a more precise way than ever before. In parallel, new ways to alter how genes are used in the body have also emerged.

Since the birth of Polly we have seen cattle able to make human blood proteins, pigs with organs customised for transplanting into people, chickens that lay eggs laced with anticancer antibodies, and goats that make spider-web silk. Despite all this promise, the full impact of this technology has yet to be felt. The techniques of genetic modification that I describe in this chapter could do even more for human medicine, a theme I shall return to at the end of the book.

Why Polly?

Our neighbouring biotech company, PPL, was keen to build on the Roslin's success with cloning by trying genetic modification, as were rivals elsewhere, notably in North America. By the time we started working on Polly, the use of genetically altered organisms – bacteria and yeast – to make proteins had become routine. These were no ordinary proteins, but ones that could be used as medicine. Among the first to be produced by yeast were human growth hormone, used to combat dwarfism; human insulin, for diabetics; and the interferons and lymphokines, such as interleukin-2, which boost the body's immune system. Called recombinant proteins, they are used extensively in the food, paper-processing and pharmaceutical industries.

However, farm animals like Polly offer many advantages over microbes when turned into living drug factories. Running bacterial fermenters – the huge stainless steel vats in which they were grown – is much more expensive than feeding sheep hay in a field. There was, moreover, a deeper problem with using microbes: these simpler organisms can make only simple proteins. More complex proteins found in people carry sugar molecules that are necessary for them to work properly in the body. Bacteria and yeast are not able to add these sugar groups, and before 'pharm' animals appeared, such complex glycoproteins had to be isolated from blood or human tissues, which was an expensive process and risked isolating human diseases too.

We were not alone in becoming excited by the possibilities of transgenic animals. The first had been created by microinjection in the 1980s, and our own efforts culminated with the unveiling of Tracy, the sheep born at the Roslin in 1990. There were many other researchers striving to become successful 'pharmers', such as James Robl at the University of Massachusetts, Amhurst, Randy Prather at the University of Missouri-Columbia, and Neal First at the University of Wisconsin, Madison. Then there were companies other than our neighbours and collaborators, PPL Therapeutics, such as Advanced Cell Technology, Alexion Pharmaceuticals, GenPharm and Genzyme Transgenics in America, and Pharming Holding in the Netherlands. If scientists could carry out precise 'genetic surgery' on an animal, the commercial potential would be huge.

From the milk of genetically altered sheep the Roslin and PPL team hoped to harvest the factor IX that Stephen Christmas lacked. Others were racing to pharm antithrombin, a protein that helps to stop blood clotting; human albumin for use in the treatment of burns; proteins that could boost the numbers of red blood cells to treat anaemia; or proteins to treat osteoporosis, arthritis, malaria and HIV. The hope was that we could make these proteins at a fraction of the cost of conventional methods. A traditional protein-drug factory would cost £100 million to £200 million and take between three and five years to build. By comparison, a new strain of sheep would cost £50 million to develop and take eighteen months to start lactating, so that we could extract drugs from their precious milk. In its lifetime one animal could make £100 million to £200 million worth of pharmaceuticals, depending on the protein.

Apart from turning sheep into 'bioreactors', as some like to call them, the introduction or deletion of genes also offered the means to make animals that grow faster for meat production, or are resistant to diseases. Bovine spongiform encephalopathy (BSE) and scrapie were obvious targets. Genetic changes could benefit the environment: one example was the 'enviro-pig', a swine that had a gene for the breakdown enzyme phytase placed in its salivary glands so that it made more use of phosphorus in feedstuffs and less phosphate ended up polluting the environment. Some researchers wanted to genetically alter animals to make them produce human milk – a rather far-fetched idea.

Several labs, including PPL, hoped that transgenic animals would eventually provide suitable organs to overcome the desperate and chronic shortage of organs for transplant into human patients. At present, thousands of people die each year before a suitable organ becomes available. PPL wanted to modify pigs so that their hearts or kidneys would work in humans. The technology could also be used to make animals designed to suffer human-like diseases for testing treatments, an application that has seen a huge surge in the use of genetically altered mice. Other companies wanted to use this method of transgenesis to enable goats to make spider-silk protein in their milk, paving the way for a new generation of wonder fabrics. The possibilities are endless.

Quite predictably, animal rights campaigners stepped up the pressure to study the welfare of transgenic animals and examine the ethics of tailoring animals for human use. We were ourselves uncertain of the long-term effects of the procedures used to create Polly. She was a first, after all. However, concerns about animal welfare should, I feel, be kept in perspective. Scientists and companies must think carefully about whether their work is justified: no one wants to modify animals for trivial reasons.

There was also unease that we had somehow 'humanised' sheep. Polly was transgenic and had a human gene in every cell of her body, after all. I should stress, though, that she was still very much a sheep: sheep are thought to carry around 25,000 genes in all, similar to the human complement of genes, and just one extra gene, even if it was 'human', would make little difference to her 'sheepness' as defined by the other 24,999 genes. Indeed, all people are somewhat sheepish. If you compare the genetic codes of human and sheep, the DNA sequences of sheep genes are about 85 to 95 per cent identical to the corresponding human genes, with the sequences of some genes more similar (scientists say 'conserved') than others. Given that a human gene itself was not used, but a clone (a copy) of that gene, Polly was a long way from being a 'humanised' sheep.

Gene Tinkering

By the time we started the Polly project, there were many ways to introduce the DNA of foreign genes into the cells of an animal. You could add calcium salts to make the DNA that codes for a gene precipitate out of a solution onto the embryo and hope that some would penetrate to the nucleus to add to the DNA already in the cell. You could use electrical pulses to punch small holes in the membranes of the embryos and drive DNA inside. You could package DNA in fat particles (liposomes), which are soluble in the greasy membrane around cells, so that they can pass inside them. The most accurate method, and also the most labour-intensive, was to inject a few hundred copies of the gene directly into the nucleus of a one-celled embryo, a zygote. This was the method the Roslin and PPL used to make their sheep. Typically, of every ten thousand embryos injected with foreign DNA, only three would make it to productive adulthood. Imagine the slog of doing something ten thousand times to achieve just three successes.

There were many reasons why this early kind of DNA modification was a lottery. Often the embryos that were injected stopped developing. Less than 20 per cent developed to term. Sometimes injected DNA failed to become incorporated into the genetic code (the genome) of the zygote until the developing embryo had formed up to eight cells. Then we would find DNA in some cells of the developing animal, but not in others – a mosaic of modified and unmodified cells, a chimera. To make matters worse, the novel DNA can be incorporated randomly in an animal's genome, and this can have unexpected effects, interfering with the action of other genes, or regions of DNA that turn genes on and off.

Even if the gene was incorporated and not causing damage, there was plenty more that could go wrong. The gene could be in the right place in the genome but not being used properly. How genes are used in the body depends on location. A gene that makes a cell hearty should be used only in the heart. Similarly, a gene that makes a nerve cell brainy should be used only in the brain. The same holds for milk-producing genes and mammary glands. There is a danger that a newly introduced gene can be turned on in the wrong tissue or organ. This is what damaged the so-called Beltsville pigs, when an American team's attempt to boost the growth of swine with human growth hormone went awry – the hormone was made and used in the wrong tissues. The result was arthritis, lameness, mammary development in males and a host of other problems. And a lot of bad publicity for this technology.

To turn a gene on in the right place – where you want it to act – you need a gene switch called a promoter. In an ideal world the promoter would be one that already controls genes in the animal you want to alter genetically. To create a pharm animal that could make the drug AAT, alpha-1-antitrypsin, in Tracy's milk, we attached the AAT gene to the promoter region of a gene responsible for a sheep milk protein, called beta lactoglobulin. The promoter ensured that the gene was used only in her mammary gland, even though every cell in Tracy's body had the AAT gene. That way her milk contained a protein that was being used to treat cystic fibrosis and emphysema, but without the expense and the risk of using AAT derived from human blood plasma.

Tracy took a lot of effort to make. This being the days before cloning, we injected a thousand embryos with this 'construct' of gene and promoter. We implanted them in around a dozen sheep a day over one breeding season. This required two people to anaesthetise the sheep, two to operate on the animal, one to remove the eggs and one to do the gene injections. This was a huge enterprise. Fortunately, it worked, and in 1990 Tracy the sheep was born. She was outstanding, a pharming pioneer who could make 30 grams of the human protein for every single litre of milk that she produced. A good dairy sheep will make 900 litres during its five-to-six month lactation period. Some breeds have three lactations in two years. That meant she could make 40,500 grams per year, a vast amount when you bear in mind that around 200 grams per year is needed to treat a single patient. In other words, Tracy could supply enough for around 200 patients all by herself.

Until Dolly was born, Tracy was the world's most famous sheep.

She was then followed by other pharm animals such as Herman the bull, created by GenPharm's Dutch subsidiary. Herman (named after a company vice-president) carried the human gene for lactoferrin, and fathered at least eight calves in 1994. Each calf inherited the gene to make this iron-containing protein, which is essential for infant growth, and the developers hoped that a new source of nutritious milk would become available. There was also Grace the goat, who produced an anticancer protein. At the time of writing, however, the first product to treat human disease to reach the market looked likely to be GTC Biotherapeutics' (formerly Genzyme Transgenics) antithrombin III agent, a blood protein with anticoagulant and anti-inflammatory properties produced in the milk of transgenic goats.

Genetic modification can be of direct benefit to farm animals. We have seen the first transgenic cows with resistance to a disease – mastitis, an infection of cows' milk glands that is often caused by the bacterium *Staphylococcus aureus*. To combat infection, a team from the US Department of Agriculture introduced a gene from the related bacterium *S. simulans* into Jersey cows which allows them to produce lysostaphin, a protein that kills *S. aureus*. In time, other strategies will emerge to provide animals with greater resistance to disease.

Polly

Polly was made in a different way from Tracy and other early pharm animals, one that promised to be much less hit and miss. My team at the Roslin – with Angelika Schnieke (the first author of the *Science* paper describing our method), Alex Kind, Karen Mycock, Angela Scott and Alan Colman of PPL – started with tissue taken from a Poll Dorset sheep. Copies of the fibroblast cells in connective tissue taken from this animal were grown in the laboratory, then human genes were introduced and labelled with a marker. Typically, these markers protected the cells from drugs that would otherwise have killed them. In this way their survival showed that the human genes had been taken up and were functioning.

In the case of Polly, we introduced the human factor IX gene along with DNA sequences that protect cells from the antibiotic neomycin. Thus, we reasoned, any cells that become resistant to neomycin after modification must also contain the human gene, a feature that could be used to identify the cells for which genetic modification had worked. This was the important step that spared us from having to transfer vast numbers of transformed embryos into ewes to see whether the genetic alteration had worked. Now we could take the nuclei of the selected cells – the parts containing the genetic instructions for the Poll Dorset sheep and the human protein – and use them to reprogram sheep's eggs from which the DNA had been removed by the nuclear transfer method. By 1997 the process of cloning had become so routine that Karen can remember little of the effort to make Polly. 'We had become a production facility and were churning out the numbers.'

The resulting embryos were transplanted into Scottish Blackface ewes. Pregnancy was easier to achieve than in Dolly's case. While only a single one of the 277 earlier attempts had succeeded, we achieved a success rate as high as one in forty-five (2.25 per cent) in the various campaigns to make Polly and her peers. This was encouraging, and suggested that the transplant of a human gene had not disrupted the development of the foetus. Whereas Tracy required around fifty ewes to create her, Polly required only around twenty. And we did not waste any resources on creating unmodified lambs – an expensive, pointless and roundabout replacement for normal procreation on the farm (although this higher initial efficiency is, of course, complicated by the reality that fewer cloned embryos are viable and only a few cloned animals end up being born).

Blood samples were taken from the resulting lambs and used to confirm the presence of the added genes. Of the five lambs, one – Polly – had the human gene plus a marker gene. The newly introduced gene enabled her to make Factor IX. PPL was obliged to tell its shareholders about this in advance of publication, so *Nature* huffily refused to carry the paper. It ended up in the equally prestigious journal *Science* towards the end of 1997. In the same issue, the birth of Dolly was celebrated as the scientific breakthrough of the year. When Polly stepped on to the scene it seemed that the earlier work with Tracy had paid off, and PPL was granted permission to assess the use of the sheep-derived human protein – AAT – to help to prevent lung damage in patients with cystic fibrosis.

The Alternative

Before the use of genetic engineering, Stephen Christmas and other haemophilia sufferers relied on blood donors to help them survive. Large volumes of blood were needed to provide the amount of factor VIII or factor IX required to control a bleeding episode, so there was always a risk of heart failure. The ability to transfuse patients with plasma instead of whole blood represented a step forward, though large volumes were still required. Then, in the mid-1960s, it was discovered that as frozen plasma was thawed a portion would remain in the form of a sludge that was rich in factor VIII (but not factor IX). This led to the development of a product known as cryoprecipitate, but it did little for Stephen.

The biggest impact on Stephen's life was to come from the supply of his missing factor in a more convenient way. Sufferers of the more common form of haemophilia were the first to benefit when factor VIII was made from large pools of plasma. Known originally as factor VIII concentrate, it was in the form of freeze-dried powder that could be reconstituted with sterile water and then injected. These concentrates revolutionised haemophilia care. Haemophiliacs were now independent of hospitals and could hope to lead normal lives. Soon, a way to isolate factor IX from plasma was also developed. Stephen could now rely on a convenient concentrate made from the plasma collected from blood donors. He had a passion for photography and became a medical photographer at the Hospital for Sick Children in Toronto, although he worked mainly as a cab driver. In this job it was easier for him not to turn up for work if he suffered bleeding into his joints.

The first step towards severing haemophiliacs' dependence on human blood came when scientists found and characterised the human genes for factor VIII and factor IX, the instructions used by the body to make these proteins. The feat was pulled off for factor IX by George Brownlee and colleagues (and others) at Oxford University in 1982 and for factor VIII, in 1984, by Ted Tuddenham's team at the Royal Free Hospital Haemophilia Centre in London, together with Dick Lawn's team at Genentech in San Francisco, and by Jay Toole's team at the Genetics Institute in Boston. These advances made it possible to use these genes in genetic engineering, and in turn to create Polly, opening up an alternative source of so-called recombinant proteins as well as making it possible to pursue novel treatments based on gene transplants.

Gene Surgery

Even the pioneering work with Polly was far from being an ideal example of how to use genetic technology to alter the genetic code of an animal. We still crossed our fingers, hoping that the human DNA we had dropped into her cells actually worked. And if it worked, we held our breath that decent amounts of protein would be made, that it worked in the right tissues, and that sticking a new gene into the genome would not disrupt the workings of other genes. This damage happened in around 7 per cent of our attempts. Even if only one of a pair of chromosomes was disrupted this way, interbreeding between two such sheep could still produce lambs in which both copies of the chromosome were damaged, along with the prospect of a healthy life.

All this uncertainty spurred efforts to refine the alteration of the genetic code of various mammals, from sheep to people. We wanted an easy method of sending genes to the right place in an animal's genetic code, rather than simply hoping that a gene could be parachuted into a creature's genetic code without mishap. One kind of genetic surgery was achieved by my colleagues at PPL in two other cloned lambs, Cupid and Diana. Employing a technique that had been used for years on mice, but with little success on other mammals, Alan Colman, Alexander Kind and the rest of the PPL team inserted a gene that coded for a human protein into sheep fibroblast cells – the flat, elongated cells found in connective tissue. They used a method called gene targeting, where they sandwiched the DNA sequence they wanted to insert between snippets of code that complement those on a target site. These 'flanking' regions of DNA exploit a built-in feature of DNA's double helix: certain letters of DNA code like to stick to each other. The trick was to match the DNA 'bookends' that surround the genetic material being inserted with the DNA on the target site. Because of the tendency for similar strands of DNA in the bookends to be attracted to those in the target genome as they entwined, there was an exchange of DNA sequences, a process called homologous recombination. In this way the researchers were able to insert a DNA sequence into a specific region on a chromosome of a sheep cell.

They chose a site for insertion that directed the production of one of the proteins that make up fibrous tissues such as cartilage, a collagen gene. They selected this gene because it is used in many different tissues and was therefore likely to be a site at which the added gene could function reliably. Even with the help of gene targeting, the gene was expected to end up in the right place only once in ten thousand attempts. As before, an antibiotic resistance gene was added too; the cells where the gene surgery had worked flawlessly could then be identified by their resistance to an antibiotic and selected for use in cloning.

Although this work was effective and took the project a step forward, since the experience with Cupid and Diana we have rethought the practical value of nuclear transfer. While fewer animals are used overall, the welfare problems linked with the present methods of nuclear transfer – to which I shall return – outweigh this advantage. We decided not to use nuclear transfer simply to add genes, reverting to the old-fashioned method of direct injection. However, cloning can be justified to change genes, since we cannot do this any other way.

You could, in principle, eliminate the kind of genes responsible for the degenerative spongiform diseases, BSE and scrapie. At the Roslin we tried to do just this. By this time, 2000, the teams had grown, with new recruits to both the embryology lab and the molecular biology lab. A team led by John Clark sought to make animals resistant to BSE in cattle or scrapie in sheep. These and other spongiform diseases result from mutations in a gene that directs the production of a specific protein within the brain, the prion protein. In sick animals this protein changes shape and triggers a molecular domino effect, converting normal prion into the abnormal form that accumulates into sheets that disrupt the function of the nerve cells. Experiments in mice showed that if the native prion protein is not produced, there can be no spongiform disease.

Inspired by this finding, we sought to knock out the prion protein in sheep. Our first objective was to prevent the gene from functioning and to show that the animals were still perfectly healthy. Then they would be exposed to abnormal prion to see whether they would succumb to scrapie. A single lamb was cloned – the first sheep with a specific gene deletion. It was not healthy, however: although it was very strong and active and had a big appetite, it panted all the time, even when resting. We did a very thorough investigation and got advice from veterinarians and even a local hospital (there are some similarities between sheep and human diseases), but to no avail. After twelve days we decided that it was kinder to put the lamb down. Detailed histopathology, carried out by Susan Rhind at the local veterinary school, revealed abnormalities in the blood vessels that supply the lungs. This was the first time that this abnormality had been seen, and it provided yet another salutary lesson that cloning a human would be very dangerous at present. Nevertheless, we had taken another significant step and there were other important objectives. Unfortunately, at this point the Geron Corporation withdrew its funding for the research project at the Roslin, and we became mere spectators of what was to follow at PPL.

Humanised Pigs

The opportunities for genetic modification have expanded as each new species has been cloned. On 5 March 2000 five piglets were born, the first swine created by nuclear transfer from adult cells. When PPL unveiled the litter, it said that nuclear transfer had proved to be more difficult in pigs than for other livestock, in part because pig reproductive biology is inherently more intractable, and in part because pigs need a minimum number of viable foetuses to maintain pregnancy (sheep, cows, mice and humans need only one). The latter problem was overcome by tricking nature: the PPL team introduced parthenogenetic ('virgin birth') embryos with the clones. The parthenotes would not develop to term, competing all the time for limited resources, but would survive long enough for the pregnancy to succeed.

The first piglet, Millie, was named after the millennium; Christa, after Christian Barnard, the surgeon who performed the first human heart transplant in 1967; Alexis and Carrel, after the transplant pioneer and Nobel prizewinner Alexis Carrel; and Dotcom to reflect the growing use of the Internet. The work was partly funded by a US government grant awarded to produce a 'knockout' pig – one that has a specific pig gene or genes rendered inactive to help prevent the human immune system from rejecting an implanted pig organ. With these piglets, PPL had shown that cloning was possible and announced at the time that it had already achieved the required targeted gene knockout in pig cells, using the same technology that led to the gene-altered lambs Cupid and Diana.

This put pigs at the heart of the effort to produce organs for xenotransplantation, the transplanting of animal organs into people. Apes would, of course, have been a closer match, but using a fellow primate would have given rise to ethical questions, and the fact that they produce only single offspring would have made efforts to humanise them hard work. Also, there would be a bigger risk that infections would spread from the animals to people – not just to the patients, but to the entire population. Their large litters and organs similar in size to human organs make pigs the most likely suppliers of any future transplant organs.

Although the size of a pig's heart is right for transplant, as is pig anatomy and blood pressure, it turns into a blackened mess in a couple of hours if human blood is pumped through it. This so-called hyperacute rejection is triggered by antibodies in human blood and then executed by a cascade of enzymes that attack the membranes of foreign cells. One pig gene in particular offered the biggest obstacle to pig–human transplants, triggering human recipients to reject the animal's organs. The gene, alpha-1,3-galactosyl transferase, is responsible for making an enzyme that adds a sugar ('alpha gal') to the surface of pig cells that is regarded by the human immune system as foreign (humans and Old World monkeys lack the enzyme and thus this sugar coating). Without the sugar, human antibodies cannot attach to the pig cells, and therefore the acute rejection process does not begin. The race was on to produce a genetically modified pig that lacked the gene.

Two years later, PPL's subsidiary in Virginia announced the birth of five healthy cloned piglets that each had one copy of the tissue rejection gene knocked out. Because the piglets were born on Christmas Day, they were given seasonal names: Noel, Angel, Star, Joy and Mary. The company thought that jolly names would deliver more column inches and make scientists look more caring and sharing than had they called their experimental subjects Porcine 1, Porcine 2, and so on. At first it was thought that PPL was the first to achieve the feat, but it turned out that a litter of humanised swine had already been produced by a rival American group consisting of Randall Prather of the University of Missouri-Columbia and Immerge BioTherapeutics, a company created by the merger of the US company BioTransplant with Imutran, the UK start-up company from Cambridge that had led the race for xenotransplantation before being bought up by the Swiss pharmaceutical giant Novartis. Their humanised swine had uttered their first squeals in September 2001, and the team had gone to the trouble of submitting its work to the journal *Science*, which obliged it to keep quiet until its work had been peer-reviewed and published.

At the time of its festive litter, PPL had announced that it expected the first application to be in the production of insulin-producing islet cells from knockout pigs for the treatment of diabetes, first in primates and soon after in humans. Clinical trials could start in four years, it said, citing analysts' belief that the market could be worth over £2.5 billion for solid organs alone and £3 billion for cellular therapies for diabetes, Parkinson's and Alzheimer's. The most headline-grabbing use of these pigs would have been for heart transplants, but there were – and still are – huge obstacles to overcome.

With the announcement of the deletion of one copy of the gene, the rival teams geared up to knock out both copies of the alpha gal gene in each pig, resulting in a 'double-knockout' animal. But it turned out that it was not necessary to make the change specifically. Spontaneous mutation in the remaining 'wild type' copy of the gene occurs frequently enough to make it possible simply to select those mutated cells. Cells in which the gene was present and functioning were killed selectively by introducing either a bacterial toxin that binds to the sugar or an antiserum that acts against the sugar residue on the cell surface. Surviving cells were used as nuclear donors and yielded pigs that lacked either copy of the gene. The two groups published their findings, PPL in 2003 and the Missouri group in 2004.

So far, so good. However, the overall objective of a double-knockout pig in itself still seemed unlikely to guarantee success: the human body rejects transplanted tissue as a result of both short-term and long-term attack by the immune system. This can lead to the build-up of deposits akin to those seen in heart disease. The details of this 'cell-mediated immunity' are poorly understood.

It remained to be determined exactly what effect gene deletion had on the pig tissues. The animals themselves are healthy enough, but what happens if blood from a human or another Old World primate is passed through the tissues? Initial results, recently published, suggest that the tissues stay healthy for a prolonged period, much longer than the few minutes before a normal pig succumbs, but much remains to be learned. Both companies claimed to have answers. Immerge said that it was looking at ways to damp rejection in patients to complement its GM efforts, a strategy that had a downside in potential side effects. PPL wanted to go further down the GM route, adding three more genes, such as DAF (decay accelerating factor), as well as genes to prevent unwanted coagulation within the blood vessels of the transplanted organ, to address other stages of rejection. But no one knew how many genetic changes were needed before a pig organ was sufficiently human for it to beat properly and pump blood in a human body without being attacked by its host. Some researchers estimated that as many as six hundred genes may be involved in tissue matching, and warned that there could be a limit to how much pig tissue can be humanised – if you altered enough of the genetic potential of the pig to make it nearly human, no one knew whether it could grow properly any more in a pig. Moreover, there was no guarantee that pig organs could do the job of a human organ for long enough to make xenotransplantation worthwhile.

A bigger issue than rejection was the fear that the use of pig organs could give rise to new diseases. In particular, scientists such as Robin Weiss of University College London worried about a group of viruses called porcine endogenous retroviruses, or PERVS. The DNA genetic codes of the viruses have over millions of years become an integral part of the pig's genetic code. These 'duplicate me' DNA messages are harmless to pigs, and not easy to spread, but critics argue that no one knew what would happen if they were introduced into humans in transplanted organs. As Weiss put it, 'Beware the Trojan Pig! When taken into your body, it may release viruses hidden inside its organs just like the Greeks in the belly of the horse.'

Intriguingly, my colleagues at the Roslin are now looking at how to introduce alpha gal into human stem cells. This is not to make them more porcine but, paradoxically, to increase the success of their transplant into patients. One fear is that if differentiated nerves, for example, are transplanted into the brain to treat Parkinson's disease, they may still contain a few embryonic cells capable of developing in an unregulated way into a teratocarcinoma, a tumour made of a blend of different cells, named from Greek words meaning roughly 'monster tumour'. (As with any therapy, there are going to be dangers as well as opportunities.) My Roslin colleague Jim McWhir has studied how to use genetic modification to ensure that alpha gal is turned *off* in differentiated cells but turned *on* in the embryonic cells. That way, the embryonic cells would be seen as foreign and rejected if accidentally grafted into a patient.

Cows, Japanese Beer and Human Blood

The going is always tough for any scientific pioneers who try to be the first to translate their bright ideas from the laboratory into the marketplace. Nexia Technologies in Montreal wanted to milk fast-maturing dwarf transgenic goats for orb spider-silk proteins. Two transgenic kids, Peter and Webster, were born in 2000 with the spider-silk gene incorporated into their genetic make-up. They became the founders of a herd that the company hoped would produce a synthetic form of spider silk called BioSteel. Bulk spider silk could be used for all sorts of applications, from body armour to medical sutures. But in May 2003 the company ran into trouble, laying off employees to save money because of delays in developing a commercial spinning process. Using the same approach, the company moved on to goats whose milk contains human butyryl-cholinesterase, a protein that seeks out and binds to nerve agents such as sarin and VX.

In 2003, PPL Therapeutics, the best known of the Roslin spin-out companies, sold its Virginia subsidiary to the University of Pittsburgh in order to stay afloat. (In 2004, PPL would leave the stock market altogether with a value of just £7 million, compared with around £500 million in the heady days when Dolly had put the company on the front pages of newspapers across the planet.) PPL's xenotransplantation technology was the most advanced of what it had to offer Pittsburgh, but it had also been attempting to create what it called an 'infectious disease platform technology': the use of GM animals to make human antibodies that are normally found in human blood, where they help to fight infection. Most therapeutic antibodies already on the market consist of a collection of identical (monoclonal) antibodies. But a real prize would be to make a collection of many different types of human antibody (polyclonal antibodies), which would have many different applications, and to do so by means of animals rather than people.

In recent years Hematech and its collaborator, the Tokyo-based Kirin Brewery Company, have managed to do just this, by combining cloning and genetic modification to alter cows – potentially a great source of these precious proteins, since a typical mature cow has over a kilogram of antibodies in its blood. Hematech in Sioux Falls, South Dakota, has unveiled a plan to develop the 'Tc cow', which it claims would be able to make human polyclonal antibodies that can help to fight antibiotic-resistant infections, fend off bioweapons such as botulinum toxin (which acts on the nerves) and anthrax, and treat immune deficiencies and other illnesses.

To create the Tc cow, the team, led by Chief Scientific Officer Jim Robl, has had to overcome several major technical hurdles. The first was how to handle the code that the body uses to make polyclonal antibodies. The region of the chromosome where all the genes that direct the production of human antibodies are located is huge, far bigger than any other sequences that have been transferred into livestock. To complicate matters, it is hard to grow cells from livestock in culture for very long, so it is even harder to carry out genetic modification and check that modification has worked than, say, with mouse embryonic stem cells, which grow almost indefinitely.

Robl came up with two clever strategies to sidestep these difficulties. First, his team assembled an artificial chromosome to carry the heavy burden of DNA required to transfer the antibody genes. Added to the chromosome were DNA sequences that offered protection against a drug. This artificial chromosome was then introduced into fibroblasts grown from a calf in the presence of the

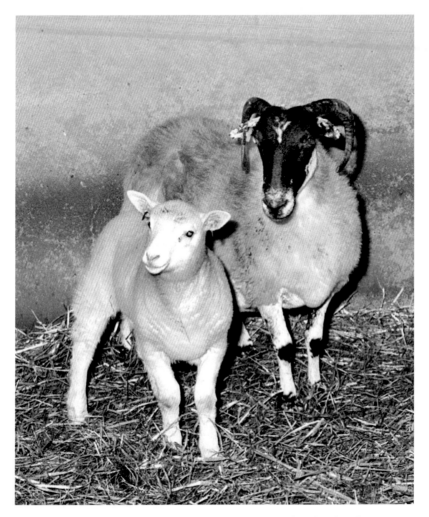

Dolly with her surrogate mother soon after her birth.
The ewe is a Scottish Blackface and obviously could not be
the natural mother of Dolly *(Roslin)*

Roslin Institute seen from the air. The building formerly occupied by PPL Therapeutics is in the foreground, the main institute is the biggest group of buildings and Dryden Farm, where Dolly was born, is in the distance *(Roslin)*

Dolly with three other lambs, all born naturally *(Roslin)*

Dolly with Harry Griffin who as Deputy Director had responsibility for public relations and both managed and participated in this work. He is seen at the time when Dolly was unveiled in the Royal Museum in Edinburgh *(Murdo MacLeod)*

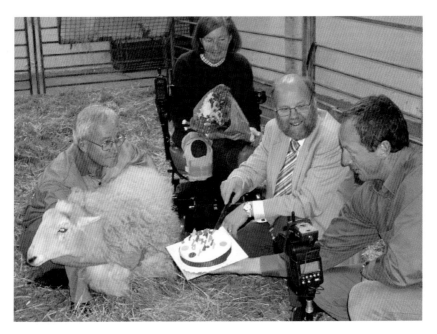

The celebration when Megan was ten years old in July 2005. She is seen with Marjorie Ritchie and John Bracken, who were involved in that experiment, and Tim King, who is now the veterinary surgeon in charge of the facility *(Roger Highfield)*

Bill Ritchie seen at his microscope with manipulator equipment *(Roslin)*

Partway through the process of enucleation the large pipette is holding the egg in place while the other passes through the zona pellucida and sucks out an area of tissue where the chromosomes are believed to be *(Roslin)*

The chamber on the microscope in which the eggs are held for manipulation. A pipette can be seen at the right *(Roslin)*

To confirm enucleation the pipette is held in UV light when DNA lights up to reveal the chromosomes from the egg and the second polar body. This enucleation was successful *(Roslin)*

The much smaller nuclear donor cell is picked up by the pipette ready for placement inside the zona next to the enucleated egg *(Roslin)*

Electric pulses are used to fuse the two cells together. This demonstration with a mouse egg shows how the two cells are arranged so that, at the point where they touch, the cell membranes are perpendicular to the electric field that flows between the two wires *(Roslin)*

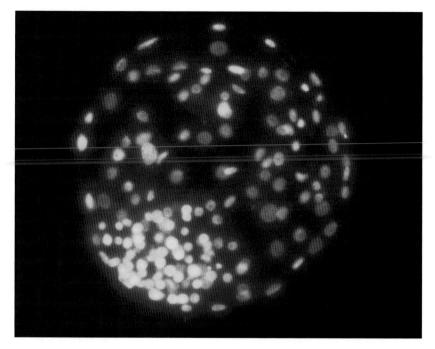

Sometimes we assess development of the eggs by staining them to reveal the nuclei, seen here as the small bright bodies. In a really good quality blastocyst there will be two hundred cells *(Roslin)*

The development of a foetus can be monitored by use of ultrasound scans, like those used with women to follow the development of their child *(Roslin)*

Dolly at maturity seen in the paddock outside her barn where she was often photographed, this time with Ian Wilmut *(Roslin)*

The Roslin Institute nuclear transfer group, including Keith Campbell (back right), Bill Ritchie (back left) and Lorraine Young (front left) *(Roslin)*

drug. Only fibroblasts that took up the protective DNA, and presumably the new chromosome, survived. However, rather than carry out a large number of time-consuming tests to confirm that the artificial chromosome was being used, the team adopted a second innovation – using the cells for nuclear transfer as early as possible, which allowed the clones to develop for fifty days before they were recovered and cells grown in culture. This offered two advantages. By the process of cloning, the nucleus was rejuvenated so that the cells lived for longer, and as a result it was possible to discover whether the cells had the additional chromosome. The cells in which the alteration had taken hold were then used for a second series of nuclear transfers, which produced the transgenic calves.

But these 'transchromic' calves were still not ready for pharming. They produced both human and bovine antibodies, and these would be difficult to separate for clinical use. To overcome this, Robl and his team wanted to use a process called sequential gene targeting, by which they could knock out both copies of the bovine immunoglobulin gene. An additional hope was that cutting the bovine antibodies circulating in the blood of the animal would significantly increase the yield of human polyclonal antibodies. That work is under way as I write. They hope that by immunising the Tc cow with a disease agent, they can create polyclonal antibodies in the bloodstream which have a strong reaction against that agent. Hematech aims to harvest antibodies from Tc cows on its production farms (each farm has around 125 cows) by a process called plasmapheresis in which whole blood is collected, the cells are separated from the liquid part of the blood (the plasma) and the cells are then returned to the body. After being collected, the plasma would be taken to a purification facility where the cow blood components (proteins, fats, potential bacteria and viruses) would be removed. The sterile, purified antibodies would then be formulated into the final product and shipped to hospitals. In 2004, Hematech and Kirin also announced they had genetically engineered a bovine embryo by removing both copies of the prion gene. The hope here, as with our earlier work, was that animals produced from these embryos would be unlikely to contract BSE. And the animals would be ideal for the production of the companies' fully human polyclonal antibodies. Pharming is now under way, not of milk or beef but of human antibodies.

Goats are being used to make another promising pharm drug, called ATryn, a human antithrombin. The drug has been developed by GTC Biotherapeutics of Framingham, Massachusetts, to treat hereditary antithrombin deficiency, a condition that makes patients vulnerable to deep-vein thrombosis – the formation of damaging blood clots. This therapeutic protein is derived from the milk of transgenic goats, selected because they are cheap, easy to look after, quicker to mature than cows and able to produce up to three kilograms of human protein annually. To create the goats, the scientists inserted the human gene for the protein into a fertilised goat's egg and, to ensure that it is activated efficiently and only in the mammary gland, added extra DNA: regulatory regions from the gene of a milk-specific protein; other DNA regions that help to maintain high-level protein production; and a piece of backbone DNA that allows the construct to be handled, replicated and injected into embryos. In this way the goats were carefully bred to maximise milk production, and thus drug production. The boss of GTC Biotherapeutics, Geoffrey Cox, says that his firm has created sixty-five potentially therapeutic proteins in the milk of its transgenic goats and cows. Given the longer-than-expected gestation time of pharming, the possibility of unforeseen problems and the exceedingly slow pace at which regulators move, it is anyone's guess when society will notice the benefits of this technology.

Gene Wizardry

In parallel with these developments, the genetic modification of animals has become easier and more efficient. One of the most important advances is a new and highly precise way to toss a spanner into the works of a cell and shut down, or 'silence', genes. It relies on new insights into the more ancient form of genetic code called ribonucleic acid (RNA), long thought to be nothing more than the dowdy ancestor of its glamorous cousin, deoxyribonucleic acid (DNA). Like DNA, RNA consists of strings of chemicals – 'letters' – that spell out the genetic code. The letters are represented by A, C, G, and either T in DNA or U in RNA. C always binds to G; A binds only to T or U. In this way a single strand of DNA or RNA can bind to another strand consisting of complementary letters, the same feature exploited by DNA targeting.

Until recently it was thought that RNA simply carries out DNA's commands, in effect translating genetic information into action. The DNA's double helix unwinds, and its genetic code is copied on to a single strand of messenger RNA. In turn, the messenger RNA shuttles the code from the heart of cells to ribosomes, the 'factories' that make the proteins that build and operate cells. But there is more to RNA than this. The first hint of RNA's versatility came some years ago, when Andrew Fire of the Carnegie Institution in Washington, DC, Craig Mello of the University of Massachusetts, and their colleagues found that double-stranded RNA can be used to shut down specific genes. Their discovery, called RNA interference, or RNAi, is now widely used as a research tool. Another clue came from David Baulcombe and Andrew Hamilton, working in the Sainsbury Laboratory in Norwich. They discovered RNA fragments called short interfering RNAs, or siRNAs, which cells use to fight viruses. Although they were working with plants, their work complemented the animal studies of Fire and Mello because siRNAs are fragments of double-stranded RNA.

RNA interference seems to occur when double-stranded RNA is chopped into siRNA. The siRNA then guides an enzyme that degrades the messenger RNA that turns each gene into a protein. This technique caused a lot of excitement because gene silencing is highly specific – the messenger RNA will be degraded only if it has the same sequence of A, C, G and U letters as the double-stranded RNA. This means that a messenger RNA, for example from a disease gene, can be targeted without the risk of side effects that would occur if the other messenger RNAs needed by a healthy cell were silenced. At Cold Spring Harbor Laboratory on Long Island, researchers have also tried using RNAs folded over like hairpins to quash specific genes. These seem to work as efficiently as siRNAs, heralding what they playfully call 'short hairpin activated gene silencing', or SHAGging. The use of RNA interference is growing rapidly and offers another useful technique for pharmers.

Down on the Pharm

We have come a long way since we tried to combine cloning and genetic modification to make the protein that Stephen Christmas lacked. Our efforts were, unfortunately, much too late to help him, though he was as influential at the end of his life as at the start, when he first gave his name to a form of haemophilia. Stephen worked for the Canadian Haemophilia Society and campaigned for the cause of blood safety, helping to secure financial compensation for patients infected with human immunodeficiency virus, HIV, despite foot-dragging by the bureaucrats and authorities.

The plasma from as many as a hundred thousand donors was combined to make blood products, such as the factor IX concentrates used to treat Stephen's haemophilia B. In the early 1980s, factor concentrates were not heat treated, because this made them less effective. In those days all it took was one HIV-infected donor to contaminate the lot. As a result, roughly 80 per cent of all treated haemophilia A patients and 50 per cent of all treated haemophilia B patients became infected with the Aids virus HIV-1.

A Canadian Haemophilia Society report, compiled largely from archival documents uncovered by Stephen Christmas, prompted the Canadian government to provide victims of the tainted-blood tragedy with compensation. Stephen learned that he was probably infected while he was sifting through recall notices for blood products in 1987, two years after the testing of blood donations for HIV and the heat treatment of blood products became mandatory in Canada. He died five days before Christmas Day, 1993, aged forty-six. His older brother, Robin, remarked on the irony that he had been killed by the factor IX that had initially given Stephen 'a shot at life'. By then, Robin had married a Jewish woman, Linda Rosenbaum, but they always celebrated Christmas in the traditional English way (Linda even learned how to make plum pudding) since they were Stephen's only close family. 'When Stephen died, so did our Christmases,' she said. 'But we always make sure to especially talk about him on that day. And we light the Jewish memorial candle of remembrance every year on the day of his death.'

In his final years, Stephen had grown bitter and angry at the authorities' indifference to the plight of infected haemophiliacs. As if to reinforce this, he died before administrative paperwork had been completed, so his estate missed out on some compensation. But his friends and family described his lifelong battle with pain as a journey of courage, learning and peace. 'He was an upbeat, well-read and well-spoken gentleman,' said one friend and fellow haemophiliac.

Our dream of using a single flock of sheep to provide the world with a safe supply of factor IX was not realised as quickly as we had hoped in the optimistic days when Polly was born. Perhaps we should have known that, like any radical new technology, it would take longer than we expected to become established. Striking the right balance between commercial hype and either social or scientific conservatism is always difficult. In the case of genetically altering animals so that their organs are suitable for xenotransplants, an impressive alternative has emerged from the very technology that helped to make it all possible: nuclear transfer for therapeutic cloning.

Rather than be used for efforts to 'humanise' pig hearts, stem cells offer a way to grow a patient's own heart muscle to carry out repairs. I remain an enthusiastic proponent of using cloning to carry out genetic alteration more accurately and safely than ever before, for instance to make animals resistant to disease. However, much more excitement has been generated in recent years by stem cell science. I myself have given up working in muddy fields and doing surgery in operating theatres for cloning human blastocysts and growing lawns of human stem cells in the laboratory.

Chapter 6

Cloning for My Father

My father, Jack, was just one person among millions who might have been helped by cells derived from a human embryo. He was diagnosed as diabetic when he was just twenty-two and finishing his mathematics degree at Magdalene College, Cambridge. He was gifted, and when he graduated in 1935 he was offered the chance to do research. But he had other ideas. He was passionate about teaching and went about inspiring the next generation with his enthusiasm for mathematics.

Jack Wilmut was not a big man, at five feet four and ten stones, but he had no problem with school discipline. He had great presence and was highly effective in the classroom. I admired him immensely. Perhaps I resemble him a bit in character; I would like to think so. Some people assume that I am a schoolteacher, and I do enjoy giving lectures at conferences, meetings and public events. My father managed to teach for only two decades, despite his ability and drive. Gradually, diabetes robbed him of his sight. His vision deteriorated to the point where it was impossible for him to teach in school. By the time he was in his forties, he was blind. Remarkably, he bore his suffering patiently and wanted to reinvent himself, this time as a computer programmer, a leading-edge profession of the day.

At that time, the British government offered retraining for another career only to younger men. My father wrote to the prime minister, Harold Wilson. He was emboldened by the knowledge that the PM owed a personal debt to the Wilmut clan: Wilson had been given remedial lessons by my grandfather, Sam, after suffering rheumatic fever as a boy and being away from school for a prolonged period. This service Wilson appreciated and never forgot. The employment rule was changed, and my father was able to learn a new skill, one better suited to his failing body. I like to believe that this was a result of Wilson's response to my father's appeal.

My father went on to work as a computer programmer for Yorkshire Imperial Metals in Leeds. But the disease continued to attack his body. He gradually lost the sensitivity in his fingertips needed to read Braille, and once again he was forced out of a job. Before he died, in 1994, he lost much of the use of his fingers, and part of one leg had to be amputated. Like any son or daughter, I would have done anything I could to reduce his suffering and make him better. If only there had been a way to halt the underlying disease by replenishing my father's body with his own islet cells, the insulin-producing cells that he lacked. One day it will be possible to do just this, by means of an embryo. From this embryo could be isolated stem cells with the potential to turn themselves into any of the two hundred different types of cell in the body.

If embryonic cells were allowed to differentiate, then from the muddle of resulting cell types the right ones could be selected to fix an ailing body. Better still, embryonic cells could be guided with growth factors down the right developmental paths to make specific kinds of cell, offering huge opportunities to repair the effects of illness, accident or disease. For curing my father, these precious embryonic stem cells could be turned into islet cells. Apart from providing cells for transplant, and a way to test diabetes drugs, this research will also provide insights into the development of the disease. And, of course, Dolly is the spiritual mother of all this work since she showed unequivocally that differentiated adult cells could be reprogrammed to be embryonic again.

For many people, the usual term for the procedure of deriving cells from cloned embryos, 'therapeutic cloning', sends shivers down the spine. Many experts in science and ethics, such as Mary Warnock, the philosopher who chaired the UK's Committee of Inquiry into Human Fertilisation and Embryology, have wondered whether the use of the term 'cloning' has damaged the field because for the general public it is so laden with grim associations and negative baggage. Understandably, the official alternative, 'cell nuclear replacement', is grey and wordy, and has not caught on. What's in a name? In this case, a great deal. These primal cells are the stuff of which medical dreams are made.

Embryonic stem cells offer the potential for making not only insulin-producing cells to treat diabetes but also nerve cells to treat degenerative diseases such as Parkinson's, heart cells called cardiomyocytes to treat damaged hearts, and other cells to fix liver damage from hepatitis or alcohol abuse. They could be turned into twitching heart cells to test new cardiac drugs, or nerve cells to rewire a broken spinal cord – you name it, it could be done. In fact, so-called cell therapy is already well established. We are all used to the idea of bone marrow transplants to treat leukaemia, and we understand that this established treatment fails if the donor's marrow is not a good match and the graft attacks its host rather than protecting it. Stem cells offer a reliable way to obtain genetically compatible tissue of any kind.

Initially, these cells have been obtained from 'spare' embryos from fertility treatments, and it is estimated that a bank of a few dozen cells from different donors could be of practical use. A perfect match with cells in this bank is possible. But even if a stem cell line produced a good match, it could still be sufficiently immunologically different from the patient's cells to cause problems. Some transplant recipients would have to take powerful and toxic anti-rejection drugs for life − drugs that increase susceptibility to infection and even cancer.

By using cloning − nuclear transfer − to grow a patient's own tissue, one can avoid rejection (there is a complication when it comes to treating the diabetes that affected my father, however, which I shall discuss later). Stem cells from a blastocyst have almost − but not quite − all of the potential of the newly fertilised egg. After starting out as the single-cell fertilised egg, the embryo divides into three different layers − ectoderm, mesoderm and endo-derm (*ecto-*, *meso-* and *endo-* come from the Greek for 'outer', 'middle' and 'inner', while *derm* means 'skin'). The ectoderm turns into nerves, skin and hair follicles; the mesoderm produces heart, blood and muscles; and the endoderm becomes the organs of the intestine, liver, pancreas and lungs. In a similar way the embryonic stem cells can turn into any of these cells.

Even in a mature adult there will always be some stem cells which are at a later stage of development than embryonic stem cells but an earlier stage of differentiation than liver, brain, bone, and so on. Another option would be to use these more developed kinds of stem cell, adult stem cells, which can develop into a smaller reper-toire of cell types for local replenishment and repair of an organ. Their versatility is still under investigation. Whereas stem cells from adult bone marrow, for example, are known to turn into blood cells, great excitement was generated by experiments that seemed to sug-gest that they were much more versatile, generating brain cells, liver cells and other types. The pro-life movement has placed great faith in these observations, claiming that they undermine the case for using embryonic stem cells. I too would welcome any use of adult stem cells, but the question remains whether the range of cells that they can be turned into is as broad as is possible with embryonic stem cells.

Austin Smith, while at the Institute of Stem Cell Research,

Edinburgh University, reported in 2002 that some adult stem cells can fuse with existing cells, creating freak cells with twice the dose of chromosomes – eighty instead of the usual forty – and unknown health consequences. In the same year, Catherine Verfaillie of the University of Minnesota Stem Cell Institute found that cells from adult bone marrow could provide an alternative to embryonic cells to make brain, heart, muscle or any other type of cells. But she herself emphasised that it was too early to say how bone marrow cells would compare with embryonic stem cells in longevity and function. It was already claimed that stem cells from marrow can make bone, cartilage and muscle, but the question was – and remains – whether these other cells can make anything beyond this expected range.

The pro-life movement has argued that there is evidence that adult stem cells, such as those from bone marrow, can transdifferentiate – be turned into nerve, heart, muscle and other types for repair – and that this makes human embryo research redundant. However, Austin Smith remains sceptical and believes that such arguments have been put forward by overenthusiastic researchers in countries where human embryonic stem cell research is severely limited. Even how to define a stem cell, let alone its potential, remains hazy at the time of writing. 'There are useful stem cell types in the adult,' says Smith, 'but the proposed paradigm shift, that blood can make brain and vice versa, has not stood up to the scientific test.'

It has been claimed that adult stem cells from bone marrow – among a wide mixture of cells – can repair a heart, and success has even been reported in patients with heart disease. But Smith argues that it is unclear whether new blood vessels and heart muscle cells really are being made from adult stem cells injected into a diseased heart. Instead, the benefit may come from growth factors and other proteins which are exuded by the transplanted cells and stimulate heart tissue repair. Not everyone is so gloomy, however.

Cell Banks

To determine which stem cells are most useful in treating a particular disease, side-by-side comparison of adult and embryonic stem cells must be done. This is one of the reasons why Britain launched a stem cell bank in 2004, the first of its kind in the world, at the National Institute for Biological Standards and Control in South Mimms, Hertfordshire, where ampoules of human stem cells of various types are laid down for storage in liquid nitrogen.

For those who regard a fertilised egg as a person, the very idea of a stem cell bank must be unsettling. Within its freezers, the genetic identity of these 'individuals' can live on almost indefinitely. The immortal nature of the cells raised 'distinct problems', according to a House of Lords Select Committee Report on stem cell research. There are have been many examples of 'immortalised' cell lines used in research about which these concerns have been expressed. Perhaps the best-known was from more than half a century ago. On 4 October 1951, Henrietta Lacks died of cervical cancer at Johns Hopkins University Hospital in Baltimore. This thirty-one-year-old mother of five might have died in obscurity had it not been for her cancer cells, now known as HeLa cells, which were the first to be discovered that thrived and multiplied outside the human body, seemingly for ever. Her cells have been used in endless experiments and even been launched into space.

Critically, for stem cells to be stored in a cell bank there should be no restrictions placed on their use by the parents of the embryo from which they came. In recognition of this and to address pro-life concerns, the British bank was established only after a lengthy process in which many legal and ethical issues were mulled over. While researchers were most interested in reliably growing stem cells into brain, nerve, muscle or any other desired types of cell, they also had to ensure that cells were grown in conditions of 'good manufacturing practice' (essentially, this means keeping a record of anything and everything that goes on in stem cell facilities) so that a convincing case could be made for the cells to be used to treat patients. Use of the cells is heavily regulated: in addition to the UK's Human Fertilisation and Embryology Authority, the Medicines and Healthcare Products Regulatory Agency has laid down conditions that have to be satisfied. The owners of any patents on processes used to create a cell line also control how the cells can be used.

Cell lines from around the world are now being sent for storage at South Mimms. They could be adult stem cells, such as bone marrow cells, cells from a foetus aborted for medical reasons, or cells from a cloned or surplus embryo, such as a blastocyst. As I have said before, the early embryo deserves respect – but so do the aims of the bankers. These precious cells will accelerate research in a field that offers new hope for a range of currently untreatable conditions. The bank will also make the most efficient use of a controversial source of tissue: no one wants to make or use any more embryos than necessary. In addition to making every embryo count, and to assist the best scientists to work on understanding these cells, the bank is critical in helping to break a monopoly of private companies that could hinder scientific progress and make stem cell treatments affordable only for the very wealthy.

Once warmed and grown at blood temperature in nutrients and growth factors, the stored stem cells can be used in experiments designed to find reliable ways of turning them into the various cell types found in the human body. After the science is perfected, a single cell can fill freezers with ready-to-use immune, brain, heart or liver cells. Turning them into tissue types will be trickier, and turning them into organs remains a distant prospect – how can one nurture them without the complex, evolving environment of a body that grows in tandem? But a ready supply of replacement cells alone could revolutionise the treatment of Alzheimer's and

Parkinson's diseases, strokes, spinal cord injuries, heart disease and diabetes.

A good match for most people could come from a bank of just 150 lines of embryonic stem cells, according to studies by Roger Pedersen's team at Cambridge, England. Pedersen left the University of California at San Francisco to take advantage of an attitude to embryonic stem cell research that is more liberal in the UK than in America. In the UK he found governmental support for such work (at the time he said that he faced a choice between 'very favourable circumstances and tremendous support' for his work in the UK and 'the prospect of sitting on [his] hands for the next few years' in the States.) At Cambridge, his team found that as few as ten carefully selected lines would be sufficient to meet the needs of 40 per cent of the population, with the aid of anti-rejection drugs.

The Great Cell Hunt

Descriptions of stem cell research in the media have created the impression that the feat of isolating human embryonic stem cells is trivial and unremarkable. It is not. The first success was reported in 1998, after six years of trying and almost two decades after a British team first carried out the same feat in mice. The first line of human embryonic stem cells was derived by James Thomson and his team at the University of Wisconsin, Madison. In November of that year, when the feat was described by his group in the journal *Science*, Thomson said, 'It's no longer in the realm of science fiction. I really believe that within my lifetime I will see diseases treated by these therapies.'

The cells were licensed to the Geron Corporation of California, evidence that even then there was intense commercial interest in their clinical uses. Only a few days later, John Gearhart of Johns Hopkins University reported success in cultivating a line of stem

cells from the germ cells of aborted foetuses. 'The potential of these
unique, versatile cells for human biologic studies and medicine is
enormous,' said Gearhart. 'These cells will rapidly let us study
human processes in a way we couldn't before. Instead of having to
rely on mice or other substitutes for human tissues, we'll have a
unique resource that we can start applying to medicine.'

Human embryonic stem cells have rarely been out of the headlines since, and the centre of gravity of the research has shifted from
the West to the East, with teams in China, Singapore and Israel
making great strides. But in Britain it took researchers five years to
catch up with the initial advances made in North America. Isolating
a line of stem cells proved extraordinarily difficult: researchers not
only needed 'green fingers' to succeed, but also had to guard against
infection, contamination or colonies morphing into mixtures of
other tissue.

Like many other groups, we study Thomson's original lines of
cells. Like many others in the UK, the group at the Roslin was also
keen to isolate the first fully characterised line of human embryonic
stem cells in the country. Various other groups are keen to have a go:
in Edinburgh, led by Austin Smith; in Sheffield, led by Peter
Andrews and Harry Moore; the team under Miodrag Stojkovic,
Majlinda Lako and Alison Murdoch at the Centre for Life,
Newcastle; and the group led by Stephen Minger – another
American – and Susan Pickering at King's College, London.

The King's team was the first to report success, in the Internet
journal *Reproductive BioMedicine Online*. Its paper described how the
King's stem cells made two molecules unique to human embryonic
stem cells, as well as others commonly found in other stem cells.
This was suggestive, though not hard evidence, that they had the
elusive cells. But they felt that their work nevertheless warranted
publication because it used embryos that had failed screening tests
for serious genetic disease (by a method called pre-implantation
genetic diagnosis, where a cell is removed from an early embryo for

analysis, which I discuss in Chapter 9). There is always a fear that, if the embryos are of poor quality in one respect or another, it may be impossible to derive a line of stem cells. The King's team considered it was worth announcing that they had helped to open up a source of stem cells that (at least to pragmatists) are more ethical to use.

The race to be the first in the UK probably ended in a dead heat, however. The Newcastle team had been growing its stem cells since the start of 2003, but delayed publishing results so that it could implant the cells into mice lacking an immune system to see whether the cells turn into teratocarcinomas, cancers that consist of a blend of all cell types. This experiment was regarded as a crucial test of whether the cells are pluripotent – capable of turning into any type. The Newcastle team, who derived the stem cells from a more conventional source – embryos left over from IVF treat-ments – did not feel that they could publish until this test had been done. The danger of a premature scramble for publicity was empha-sised by an earlier false dawn in British stem cell science. In August 2002, an overenthusiastic BBC news programme reported that King's had pulled off the feat. That cell line was subsequently lost to contamination.

Roslin's Stem Cells

The Roslin's stem cell effort began in earnest in May 1999, when the institute signed a deal for research on human embryonic stem cells with the Californian company Geron. As part of this pro-gramme we set up a 'clinical grade' laboratory for the isolation and growth of stem cells, led by Paul De Sousa and backed by Scottish Enterprise and the Medical Research Council. Like other teams, we were excited by the thought of answering a very simple question: if nature can direct a single, non-specialist 'über-cell' – the embryonic stem cell – to develop into masses of specialist cells, such as bone, heart, muscle, and so on, then why can't we?

We wanted to extract the essence of 'stemness' and discover what allows stem cells to specialise and become a member of a tribe, whether the skin tribe or the eye tribe. Several factors were known to influence how to turn a stem cell into an endoderm cell, a cell that could help rebuild the pancreas of a diabetic patient such as my father. The challenge facing my team and all the other teams is to decipher all the components of the complex chemical environment within the body that reliably determine a stem cell's fate.

It is estimated there must be about twenty pathways in the body governed by hormones and growth factors that control how a stem cell turns into a particular cell type. Like the other teams, we are looking at how to tweak all twenty in a systematic way by the use of automated methods. For example, it is possible to put the cells into an array of a hundred or so tiny wells arranged on a microdish, and then use robots to move between the wells and alter conditions one by one so that experiments can be conducted on all of them simultaneously. One could arrange for there to be a gradient of concentrations, from high to low, in each of two different chemicals in the recipe, from left to right and top to bottom in the array, in order to test the influence of both concentrations on the ease with which one can incubate, grow and differentiate the cells. By using a hundred microdishes, one could look at ten thousand combinations of conditions, and in a few weeks complete a set of experiments that would keep a conventional laboratory tied up for a year. As a bonus, these microdishes work with a minimal number of cells, growth factors and other precious substances (human proteins cost a fortune).

By finding a precise recipe that can reliably turn stem cells into the desired type, we also cut the risk of contracting animal or human diseases. The traditional way of growing human embryonic stem cells is to nurture them on animal (usually mouse) 'feeder' cells in a broth – which is as unscientific as it sounds – consisting of foetal calf serum, a complex mix of substances with an action that is not fully understood.

In 2005 the risks of this approach were underlined when contamination was found in the 'presidential' lines of human embryonic stem cells – President George W. Bush had in 2001 restricted federal funding for the development of new treatments to the limited numbers of embryonic stem cell lines available at that time, a move that annoyed pragmatists and fundamentalists alike. Ajit Varki, of the University of California, San Diego, and Fred Gage, of the Salk Institute, found that the lines contain a foreign substance, N-glycolylneuraminic acid, and suggested that if they were introduced into patients there was a significant chance of an immune reaction and/or rejection of the transplanted cells.

Since this setback, new cell lines are being derived at the Roslin by Paul De Sousa, backed by Geron. They have been grown with the help of fibroblasts from human foreskin and human laminin, from placentas, but the hope is that one day we can have the equivalent for human cells of hydroponics for plants, where soil is replaced by a solution of nutrients. We wanted to get rid of feeder cells altogether so that we knew exactly what our stem cells were dining on and could ensure that this recipe would produce cells of the desired type. We have to get this recipe absolutely right if these cells are going to be safe to use. If a stem cell runs amok, the result can be the type of cancer called a teratocarcinoma, a grotesque tribute to the stem cell's flexibility. The cancer is a mixture of gut, muscle, nerves, teeth, tufts of hair – even facsimiles of fingers and legs – that I mentioned earlier as one of the tests of pluripotence. We have to make sure that the cure is not worse than the disease.

Eventually, with colleagues at the University of Edinburgh, Jim McWhir hopes to turn stem cells into bone cells and cartilage, while Wei Cui's team at the Roslin is studying how to make liver cells for drug testing (since the liver enzyme cytochrome p450 breaks down most drugs) and the right kind of brain cells to treat Parkinson's disease. As well as using established lines of human stem cells, my colleagues are trying to create new lines in these defined culture conditions. One source of new lines is blastocysts from IVF clinics which are either surplus to requirements or have failed genetic screening for congenital diseases and would otherwise have been discarded. Instead, and with the permission of the parents, they can be used in experiments. Most people would accept that this is an ethical source of cells, particularly given the usual fate of these blastocysts.

Virgin Cells

There is, however, an alternative source of human embryonic stem cells that does not – or at least should not – count as a human life. The human embryo in question has grown not from a fertilised egg but from an egg that has been tricked into dividing. It is as though it were possible, for a short while, to rekindle some of the ancient chemical machinery that enabled single-celled creatures to clone themselves. Paul De Sousa and I have been given permission by the HFEA to pursue an idea that goes back to the very start of life on Earth and gets its name from the ancient Greeks: parthenogenesis, which literally means 'virgin birth'. One early reference to parthenogenesis can be found in Greek mythology, when Athena emerged from the forehead of Zeus (Athena's temple on the Acropolis is called the Parthenon for this very reason).

Most plants can multiply by parthenogenesis, and so do all fungi and many animals, such as corals. Other examples range from aphids and bees, the male drones of which come from unfertilised eggs, to the timber rattlesnake and the basilisk, called the 'Jesus Christ lizard' for its ability to walk on water. Even the turkey can do it. But parthenogenesis cannot naturally give rise to offspring in mammals, such as humans. The reason has to do with a process called imprinting in which the mother and father have a say in how genes are used.

Any attempt to persuade an unfertilised mammalian egg to develop to term is doomed by imprinting, a phenomenon I shall explain in detail in Chapter 8. Although you inherit two copies of every gene, one from each parent, only one is used in the body. Imprinting governs which is used – the one from the mother or the one from the father. In this way, mums and dads shape how their genetic inheritance is spent. In mammals, at least seventy-five genes (probably many more) with diverse and important functions during development are thought to be regulated in this way, so they must be inherited from the right parent for a pregnancy to succeed.

Imprints are critical for the creation of the placenta, the embryo's lifeline to the mother, and the disruption of imprinting of other genes is linked with schizophrenia, immune deficiency and forms of cancer. Imprinting also explains the cause of Prader–Willi and Angelman syndromes, marked by mental retardation and a host of problems: they strike when a baby fails to inherit a paternal copy of chromosome 15, or fails to inherit a maternal copy, respectively. And imprinting explains why there are no virgin births: the embryo cannot develop without a genetic contribution from the father. At least not to term.

The Mouse with Two Mothers

We started our parthenogenesis effort at the Roslin before the birth of a remarkable rodent that highlights the importance of imprinting, how it arose and also how to defy it. Joining Dolly the sheep on the list of headline-grabbing creatures that have pushed forward the frontiers of science is Kaguya the mouse. Although she supposedly came into this world through parthenogenesis, at least according to the journal that announced the feat, scientists have argued about what to call a mouse that has two mothers rather than just one.

Kaguya leapt over the apparently huge obstacle of imprinting that prevents a mammalian parthenogentic embryo – a parthenote – from developing. She was born in 2004 as a result of work by Tomohiro Kono of the Tokyo University of Agriculture, with colleagues in Japan and Korea. Rather than combine the imprinted male and female genetic codes, as normally happens after egg meets sperm, Kono combined a female genetic code from an older egg with a 'male-like' female genetic code created from the genetic make-up of a young egg, which is free of most imprinting. To masculinise the imprint-free female code he used genetic engineering to remove a gene called H19 from the young egg so that it was being used only in the old egg. As a result of the removal of this gene, production started of a nearby (in the sense of the genome) growth factor, Igf 2, that is normally made only from the male genetic code. This growth factor plays a key role in the development of the placenta and other tissues. The resulting first mammal with two mothers, named Kaguya after a Japanese fairy-tale princess, grew to be a healthy adult female that was able to produce offspring in the conventional way.

Kaguya provides backing for a theory that explains why the peculiar business of imprinting evolved in mammals. Put forward by David Haig of Harvard University and his colleagues, it rests on the most fundamental feature of all life: the urge to propagate one's genes in future generations. Imprinting results from an evolutionary battle of the sexes, reflecting the different genetic agenda of each parent, according to Haig. To put it crudely, both parents want to pass on their genes to their offspring while still fulfilling the selfish urge to 'duplicate me'. Conflict arises because a woman's genes also have her interests – not the man's – at heart, and vice versa.

The father wants his offspring to grow big. This will provide his offspring – and his genes – with a better chance of survival and tie up the mother's resources so that they won't be wasted on another man's offspring. Women, on the other hand, have to share their food and other resources with the unborn child and reduce the risk posed by excessive foetal growth, which would be deadly. The mother also wants to survive to pass on her genes to future children. The result is an evolutionary arm-wrestling match, with paternal genes striving to beef up the offspring and maternal genes holding growth in check. Parthenogenetic human and mouse embryos fail to develop because they contain a double copy of growth-limiting maternal genes, but no paternal rivals such as Igf2 to boost growth. Kaguya overcame this obstacle, but it is still a mystery how the Tokyo team's simple genetic modification managed it, albeit in only one out of 450 or so attempts.

In our experiments we were not trying to make a human Kaguya. We thought that human parthenotes could aid our work on therapeutic cloning while sidestepping some of the ethical issues. And we did not need to use genetic modification, because imprints seem to have the most influence later in development, making it possible to defy them for a few days. A human egg can be persuaded to divide with a shock of electricity or a dose of chemicals. The unfertilised egg can also be tricked by chemicals into keeping the set of chromosomes it would otherwise have jettisoned in polar bodies. (In normal sexual reproduction, an egg loses half of its chromosomes, so when it docks with sperm during fertilisation the full number is restored in the resulting embryo.)

The result is a parthenote which retains the full set of DNA of one woman. Although it cannot develop to form an offspring – indeed it does not develop for more than a few days – the parthenote can provide insights into imprinting and act as a source of embryos which can be used to learn how to derive stem cells. It is not certain whether such cells could be used to treat patients because of the errors that occur in imprinting. However, if these errors do not interfere, cells from parthenotes could be used to treat the woman who had donated the egg for parthenogenesis. During the lengthy process of egg development (meiosis) the genes of a woman's mother and father are scrambled by a process called recombination, so her eggs have a different genetic make-up. Still, they should not 'look' foreign to the egg donor's immune system, and the cells could be of value for treatments for which an exact tissue match is probably not required, such as for the creation of nerve cells for transplant into the brain to treat Parkinson's (the brain is 'immune privileged' and thus more tolerant of foreign tissue). Parthenogenesis also offers the opportunity to create a cell line from a woman who has a serious genetic disease, so that a very detailed study can be made of the cellular effects of the disease.

The parthenogenesis programme also helped with the most controversial part of our embryonic stem cell research at the Roslin: the use of nuclear transfer to create embryos. Because part of the nuclear transfer method involves using electricity or chemicals to persuade a reconstructed egg to divide, the similar methods – the 'activation protocol' – used to trick an egg into dividing to form a parthenote provide vital information on the best way to do this and to launch development.

Never Say Never. Again

The same year we were granted a licence to create human parthenotes, some scientists began to suspect that humans might be particularly hard to clone. They had, of course, forgotten the central lesson of recent reproductive science: never say never. In 2003 Gerald Schatten, director of the Pittsburgh Development Center, observed that there were what he called 'molecular obstacles' that blocked normal cell development in cloned rhesus macaque embryos. His team tried four different cloning methods on 724 eggs retrieved from female rhesus macaques. In the resulting embryos the chromosomes did not separate properly when cells divided.

A fundamental process in cell division is the pulling apart and segregation of chromosomes equally into the two daughter nuclei; to do this, the cell draws upon the cytoplasm to build a threadlike structure called the spindle to precisely align and separate the chromosomes. An abnormality in the spindles of human embryos before implantation, caused by abnormal replication of centrosomes (the organising centres and precursors of spindles in the cell), is thought to be one reason for many of the chromosome defects observed in early human development.

As a result of cloning, however, Schatten found 'chaotic' spindle structures and unequal chromosome counts. Even the most basic proteins involved in spindle formation were absent or inadequate. Schatten concluded that when it came to primates and people, therapeutic cloning would be difficult, and 'reproductive cloning' – the creation of cloned babies – almost impossible. This seemed too pessimistic. Perhaps in non-human primates the normal procedures for removing the genetic information from the egg (enucleation) also extruded something important from the cytoplasm. Perhaps a protein that plays a key part in the formation of the spindle is taken away too, wrecking the prospects for development. As ever, small changes to the basic cloning method could overcome this. For example, the donor nucleus could be put into the egg before removing the egg's genes. An article in the journal *Science* concluded, 'There are potential ways around the newfound obstacle, but for now, groups that made controversial claims that they would use the techniques that produced Dolly the sheep to create human babies are unlikely to succeed.'

Given the pressures on the American scientific community from the Christian Right, one cannot help wondering whether some scientists yearned for a technical roadblock that could somehow end all hopes of reproductive cloning, yet keep the vast promise of therapeutic cloning alive. The echoes of Solter's gloomy pronouncement on the prospects of cloning mammals were unmistakable. There were also more distant echoes of that age-old human desire to feel special, unique and somehow separate from sheep, amphibians and the rest of the animal kingdom.

The Lords, Regulators and Roslin

The year before Schatten published research that cast doubt on the viability of human cloning, I announced that the Roslin would apply, together with Christopher Shaw from the Institute of Psychiatry, London, for what was then widely expected to be Britain's first licence to clone human embryos. I first discussed it in public in 2002, at a conference in Berlin. Although cloning human babies is outlawed in Britain, therapeutic cloning was made legal that year, after the House of Commons and the House of Lords both voted it through with clear majorities. Little did we realise that the Roslin project would face many delays. The first submission to the Human Fertilisation and Embryology Authority was to come from another British group, while the first (apparently) convincing cloned human embryos were to be produced on the other side of the world.

First, though, we had to wait for the dust to settle from legal submissions to the HFEA by pro-life campaigners. This setback was entirely acceptable, since it was essential to establish a clear legal path for the research. Another cause of delay was the HFEA's adherence to the recommendations of a Lords Select Committee stipulating that embryos should not be cloned for research unless there is an exceptional need that cannot be met by surplus embryos. The implication that a specific justification for cloning embryos is required appears to be at odds with the admission by the Lords that cloned embryos are useful for research.

The project slipped further when I became bogged down in paperwork, one of the greatest problems facing researchers in Britain today (although I fully accept that some paperwork is needed for such complex issues with major social implications). In order to obtain permission to do research on cloned human embryos, it is necessary to submit the proposal and all the associated forms to ethics committees at the place of the research and at any hospital that may supply eggs or cells for use in nuclear transfer. The proposal is then assessed for scientific validity and ethical acceptability. Having cleared the local hurdles, one then has to send the forms to the HFEA, which circulates them to outside referees for scientific, clinical and ethical assessment. The applicants get the opportunity to respond to any comments. These comments are then submitted to a final licensing committee. The first, informal contact between our group and the HFEA was on 11 October 2001, but it was not until 8 February 2005 that our licence was granted. Many people, myself included, contributed to this delay, but it would surely bring despair to patients who hope to benefit from research in this area.

Our application was about creating entirely new opportunities to study motor neuron disease, a debilitating, degenerative illness – actually a group of related diseases affecting the 'motor' (movement) neurons in the brain and spinal cord. In some countries it is also known as amyotrophic lateral sclerosis (ALS) or Lou Gehrig's disease, after the famous baseball player who was a victim. These neurons are the nerve cells along which the brain sends instructions, in the form of electrical impulses, to the muscles and extremities over distances of up to a metre. Degeneration of the motor neurons leads to muscle weakness and wasting. It first strikes the arms or legs, some groups of muscles being affected more than others. Some people may develop weakness and wasting in the muscles supplying the face and throat, causing problems with speech and difficulty chewing and swallowing. Eventually, breathing is affected.

We have no understanding of the genetics of the disease, or of the extent to which various nerve cell types are affected. Mutations in one gene, called SOD-1, have been linked with the disease: there is an accumulation of the abnormal SOD-1 protein in cells, as occurs in other diseases such as Parkinson's and Alzheimer's. But this gene is responsible for only 2 per cent of all cases. The mutations responsible for the other 8 per cent of inherited cases have not been found. In 90 per cent of the cases, genes also play a role, albeit a weaker one, but these cases are 'sporadic' and their development is thought to be influenced by environmental factors.

More than four thousand people are affected with motor neuron disease in Britain alone, and a thousand die every year, usually of respiratory failure. The disease can sometimes take longer than a year to diagnose. No effective treatments exist, though more progress has been made in the past decade than in the preceding century. To boost the prospects for new treatments and methods of diagnosis, we wanted to collaborate with the Institute of Psychiatry in London and use nuclear transfer to create human blastocysts to derive normal and affected stem cell lines.

We intended to hone methods to create motor neurons reliably from these blastocysts. The idea was to create at least one normal motor neuron cell line, one with the SOD-1 mutation, and others from patients with another inherited form of motor neuron disease to allow comparisons to be made. We sought to pin down exactly the changes that occur, particularly in the earliest stages. We hoped to lay bare how the disease affects gene use, protein manufacture and cell metabolism. We wanted to find exactly what causes a healthy motor neuron to decay and die.

We could see other benefits springing from this research: a line of nerve cells that suffer from motor neuron disease could help in developing better tests for the disease, so that it could be diagnosed

early enough to give any forthcoming treatments a chance to
work. They could be used to screen any new drugs under development; up to a hundred thousand potential drugs could be tested
each year at a total cost of just £20,000. This is a bargain in terms
of money and time: testing a single drug in clinical trails on five
hundred patients takes two to three years and can cost £10 million.
In this way, a line of affected nerve cells offered an alternative to
animal testing, although opponents of any kind of cloning would,
of course, find the idea of 'embryos for medical testing' as objec-
tionable as 'embryos for spare parts'. In the long term, stem cells
may offer a new kind of treatment in their own right: experiments
on rodents suggested that stem cell transplants into the spinal cord
could help prevent decline. The implanted stem cells manufacture
cell survival factors that help their neighbouring cells and in themselves help repair damage. Hans Keirstead and his colleagues at the
University of California, Irvine, for example, have shown how to
obtain from human embryonic stem cells almost pure populations
of the specialised cell type that insulates nerves (oligodendrocytes).
Injections of these cells improves mobility in rats with spinal cord
injuries, providing evidence that these cells can help to restore
motor skills lost from acute spinal cord tissue damage.

The First Cloned Human Stem Cells

While we were wrangling with the HFEA over permission to start
our efforts to clone human blastocysts, we were overtaken by other
groups. The first supposedly convincing evidence that a stem cell
line had been produced from a cloned human embryo came from
a South Korean team. Before their work was published in 2004,
claims to have created a human clone had been made on several
occasions. This time, however, details of how to do it were published
in one of the world's most respected science journals – and by a

reputable team, led by Woo-Suk Hwang, not, as occasionally was the case, by a cult, a self-publicist or a maverick scientist. And the human clones that the team produced were much more advanced than those from other earlier attempts by reputable scientists, having been grown to the blastocyst stage.

Other teams had tried to clone human embryos and found that they failed to develop properly, only managing a few cell divisions. In *e-biomed: The Journal of Regenerative Medicine*, the American company Advanced Cell Technology (ACT) claimed to have made at least one six-cell embryo by means of the Dolly technique. But at five days old this embryo consisted of far fewer than the hundred or so cells that one would expect it to have, and was probably dead. Nor was there even evidence that the DNA in the clone had come from the donor cell.

The lack of truly novel science in the ACT paper prompted a walk-out on the editorial board of the journal by Robin Lovell-Badge, a highly respected stem cell scientist from the National Institute of Medical Research at Mill Hill, London, and two figures familiar from earlier chapters, Davor Solter and John Gearhart. It was not that the work was wrong or because it was fraudulent, but because it did not represent an advance. 'There was no scientific reason for publishing this. I could not believe that the paper had been refereed properly,' Lovell-Badge said at the time. Gearhart complained that the announcement was aimed at investors and the public but not at scientists. 'I doubt that these scientists have any problem with the data in the paper,' responded Michael West, president and CEO of Advanced Cell Technology. 'As I understand it, they disagree that such preliminary results should have been published. On that issue we simply disagree. Our belief is that the data on this research needed and continues to need absolute transparency with the public.'

In 1998, a Korean group claimed that it had cloned a human embryo by nuclear transfer, but its experiment was terminated at

the four-cell stage, before the DNA is put to work in the developing embryo, so they had no evidence of successful reprogramming
by nuclear transfer. Then there was the curious case of the Chinese
team who claimed to have succeeded in putting human DNA into
rabbit eggs, though again some scientists were sceptical. Similar
independent studies in Japan have achieved development to the
blastocyst stage after transferring monkey nuclei into rabbit eggs.

Despite Gerald Schatten's fear that humans were particularly hard
to clone, the obstacles were overcome by Woo-Suk Hwang and colleagues in South Korea. Details of his experiments were published
in the journal *Science*, and Hwang, with his colleague Shin-Yong
Moon, presented the findings at a feverish press conference held in
February 2004 at the annual meeting of the American Association
for the Advancement of Science, held in Seattle. Hwang read to the
journalists from a written note which declared, 'Our goal is not to
clone humans.' Co-author Jose Cibelli of Michigan State
University, who had made intellectual contributions to the manuscript, along with a genetic analysis of non-human primate cells (all
human experiments were done in Korea), remarked at the time:
'When I first saw their manuscript, I almost passed out. This was it.
This paper proves that you can take a cell committed to being one
type of tissue and make it go back in development. This is the first
time it was done in humans.'

Korea had a powerful stem cell research effort, with an estimated
four hundred scientists taking part. While Hwang was based in
Building no. 85 at Seoul National University, Moon, a gynaecologist,
worked in the Seoul National University Hospital, on the other side
of the city. The two labs received most of their funding from the
South Korean government – a relatively modest £5 million a year.

Hwang grew up in Chungcheong Province, three hours from
Seoul, amid the poverty of the Korean War and its aftermath. He
studied veterinary science, and made his reputation investigating the
reproductive biology of cattle, moving to human material after the

American stem cell breakthrough in 1998 (still working on pigs for xenotransplantation and on BSE-resistant cattle). Hwang's team seemed to have produced more human embryonic stem cells – most of them not cloned – than any rival team elsewhere in the world. A science superstar in his country, Hwang was awarded a decade of free travel for himself and his wife by the nation's flagship airline. Domestic media lionised him as a scientist with 'God's hand'.

For their first human cloning study, Hwang's team collected 242 eggs from sixteen unpaid women volunteers (this would later be questioned). In a laboratory where the patient grind of cell surgery was combined with automobile assembly line methods, scientists dressed in blue jumpsuits, masks and bonnets worked on micromanipulators to transfer into those eggs the DNA from the women's own cumulus ('cloud') cells, which surround the egg and once helped nurture the egg's development. The team was dedicated and worked almost every day of the year. When human eggs were not available they practised with cattle eggs from the slaughterhouse. The team used a chemical to trigger a surge of calcium in the egg, mimicking what happens on fertilisation by sperm, making the cloned embryo divide and start development.

By tweaking the time between the transfer of the nucleus and the activation of transplanted DNA, Hwang's team was able to optimise its results: a two-hour delay seemed to work best, with 20 per cent of all reconstructed eggs then forming blastocysts. The group had thirty successes and attributed them to stringent timing and the use of extremely fresh donor eggs – offering a huge advantage over the use of embryos that are surplus to, or rejected from, IVF – and a special method for gently extruding rather than sucking DNA from eggs.

The human blastocysts were cultured for a few days until they consisted of one hundred cells, of which twenty-five were the inner cell mass, the embryo proper. From twenty suitable inner cell masses, the Korean team managed to obtain only a single human embryonic stem cell line – and thus identical tissue for one woman donor. To demonstrate the versatility of these much sought-after cells, the team found that this line could turn into three of the main tissue types – ectoderm, mesoderm and endoderm – that appear at the beginning of development, when the embryo is called a gastrula and organised into three layers.

When transplanted into mice, the stem cells differentiated into still more specific cell types, offering further proof that they were stem cells. The team could verify that the embryonic stem cell line was genetically identical to the donor of the cell used for cloning by carrying out DNA fingerprinting analysis, and to make sure that it was not the product of parthenogenesis, they could check the use of imprinted genes. But they only succeeded in cloning when using eggs and cells from the same woman.

There were other qualms about the initial Korean work. They had managed to obtain 242 eggs, free, from volunteer donors. This raised the eyebrows of American rivals, who would have had to pay thousands of dollars. 'Our ethics board never would have allowed that,' said one. 'They told us if you don't pay them anything it's exploitation, and if you pay too much it's coercion.' Some ethicists wanted to know more about how the South Korean team attracted the sixteen volunteers. Hwang offered an explanation: 'Some young ladies have a lot of curiosity about reproductive cloning and therapeutic cloning, and after searching the website they contacted us.' At the time, Hwang vehemently denied that any women in his team donated the eggs. And he stressed that his colleagues had indeed followed guidelines.

Any lingering doubts about the team's efforts seemed to be swept away a year later, when they refined the technique and showed for the first time that it could be used to combat disease. In another paper published in *Science*, Hwang (and twenty-four co-authors) described how eighteen women donated 185 eggs and how, with cells from donors, they were used to clone embryos and then create lines of embryonic stem cells. Predictably, the American president voiced his concerns about where this work was leading: 'I worry about a world in which cloning becomes accepted.'

The cells to reprogram the 185 eggs had been obtained from eleven donors, males and females ranging in age from two to fifty-six, including individuals who had juvenile diabetes, spinal cord injury and a genetic immune deficiency called congenital hypo-gammaglobulinaemia. This time there were no quibbles about consent. Nuclear transfer was used to create thirty-one embryos which were used to create eleven lines of human embryonic stem cells, marking an average of seventeen eggs for each stem cell line. For egg donors aged under thirty, an average of fewer than fourteen eggs was needed to generate a stem cell line, a stunning level of effi-ciency – sixteen times better than before. There was also evidence that the cell lines matched the patients' cells and did not have a parthenogenetic origin, when unfertilised eggs divide on their own.

Success depended on many factors, said Gerald Schatten, co-author of the Korean work: human feeder cells – not animal cells – were used to grow the embryonic cells; Koreans, experienced at handling steel chopsticks, seem more adept at micromanipulation of eggs and embryos; the eggs were from young donors, of high qual-ity, fresh and handled more gently; the team avoided the use of antibodies and chemicals to obtain cells from the embryos; and the donor DNA was retrieved with a light touch. Hwang himself con-siders that it is also significant that at least one of his researchers keeps the precious cells company all day and most of the night, as a way of nurturing respect for them. 'You need the heart and the spirit, the human touch,' he said.

Scandal

Hwang was by now so popular that he would be mobbed if he went out to eat at a restaurant. The veterinarian joked that he had become Korea's answer to Elvis. He was indeed a rock star of science, one even complete with an online fan club entitled 'I love Hwang Woo-Suk'. The poor boy from a desolate village who had risen to become a giant of cloning by working seven days a week, fifty-two weeks a year (his motto: 'No Saturday. No Sunday. No vacations'), had inspired at least ten popular books in Korea.

This source of immense national pride was about to become a national embarrassment. Triggered by a report in the journal *Nature*, there had been concerns about exactly how Hwang had obtained human eggs for his research. In the closing months of 2005, the matter came to a head when Gerald Schatten announced he would suspend his collaboration because of 'misrepresentations' about egg collection 'and the resultant breach of trust'. Days later, in an emotional press conference held in Seoul, Hwang admitted that two female scientists in his lab had given their own eggs for research. Under commonly observed international guidelines, this is prohibited so that vulnerable scientists cannot be put under pressure to donate in what is a painful and sometimes risky procedure. Hwang apologised profusely, confessing he had lied when the claim was first made in *Nature*. 'I am very sorry that I have to tell the public words that are too shameful and horrible,' he said.

The 242 eggs that Hwang had used to create the first human cloned embryo had come from sixteen women, said by him to be unpaid donors. The Health Ministry confirmed that one of the doctors (later revealed to be co-author Roh Sung Il of MizMedi Hospital) who had earlier been collecting eggs for Hwang's research paid thousands of pounds to some women for their eggs, which was not then illegal in Korea, but is now. And it said that an investigation found that two junior scientists had given their own eggs for research, confirming that the donations weren't in violation of the then current ethics guidelines because they were made voluntarily. Hwang still maintained, however, that when two researchers offered their eggs he had turned them away. According to him, in 2003 they went behind his back and donated eggs under false names. He added that he had publicly denied reports that this had happened because one researcher asked for her privacy to be respected.

South Koreans rushed to support their science hero. Protesters picketed the headquarters of South Korean broadcaster MBC after the network aired a TV programme that focused on the egg donations by the junior researchers. There was a jump in the number of South Korean women volunteering to give their own eggs for Hwang's work, with hundreds more coming forward. Women even left bouquets of the national flower, hibiscus, at his laboratory, along with touching notes of support.

The hero retreated to a remote Buddhist temple, then was admitted to hospital suffering from exhaustion. Slowly he recovered. But when a weary Hwang returned to his lab early one Monday morning, after an eighteen-day absence, the scandal was far from over. Schatten had by then asked the journal *Science* to take his name off the cloning study published by Hwang and colleagues in May 2005 in which he described how he had used nuclear transfer to make human stem cell lines. 'My careful re-evaluations of published figures and tables, along with new problematic information, now cast substantial doubts [on] the paper's accuracy,' Schatten said in a letter to *Science*. 'Over the weekend, I received allegations from someone involved with the experiments that certain elements of the report may be fabricated.'

Internet sites buzzed with allegations about the 2005 paper. One question was whether photographs, described as showing stem cells derived from cloned human embryos, were frauds. A co-author, Roh Sung Il, said that they were not derived from cloning experiments. Another question concerned the veracity of the DNA fingerprints used to show that a stem cell was genetically identical to a person who provided cells for cloning. They were 'too clean', said one researcher. Another talked of 'unusual similarities'. However, in a nationally televised news conference,

Hwang said that the data were essentially correct and that he stood by his findings.

I had been through a similar experience with Dolly and I knew what had to be done. I added my signature to those on a letter asking Hwang to submit his work for independent scientific confirmation. By now the unfolding saga had triggered anger and bewilderment among the public. But there was still a bedrock of support. In mid-December the 'I love Hwang Woo-Suk' website declared: 'No matter what anyone says . . . our love for Professor Hwang Woo-Suk will never diminish.'

Even then, I too was still convinced that Hwang's team had indeed succeeded in cloning human embryos. I had visited his laboratory and been impressed by the scale of his effort, which had been backed by more than £20 million from the South Korean government. I sympathised with his dilemma: I myself had to do follow-up work to convince high-profile sceptics in America that Dolly was indeed a clone of an adult cell. I also know only too well how media reporting can spin out of control. Then I found out that in January of 2005 Hwang had reported to his government sponsors that six of his stem lines had been lost to fungal contamination, an important fact not mentioned in his report later published in *Science*. I felt uneasy.

An expert team from Seoul National University looked into the disputed paper. On 23 December, the investigation delivered its preliminary conclusions. I was on holiday at the time but the shattering findings were inescapable because they made headlines worldwide: the data for the eleven collections of stem cells Hwang claimed to have made were derived from just two stem cell lines. 'Based on these facts, the data in the 2005 *Science* paper cannot be some error from a simple mistake, but can only be seen as a deliberate fabrication,' the panel said.

The team found no records of two of the other stem cell lines. Four others had died from contamination, and another three hadn't yet become full stem cell lines. To create fake DNA results purporting to show a match between the clones and patients, the panel claimed that Hwang's team split cells from one patient into two test tubes for the analysis – rather than actually match cloned cells to a patient's original cells.

Also in question was Hwang's remarkable efficiency, by which he managed to use just 185 human eggs to create custom-made embryonic stem cells for the eleven patients. The investigation had, said SNU professor Jung-Hye Roe, 'found that there have been a lot more eggs used than were reported', suggesting lower efficiency (the investigation would conclude that between November 2002 and November 2005 the team used 2061 eggs from 129 women – one of the more shocking statistics given the risks these women had taken in being stimulated by powerful hormones (which can be life threatening in 1 or 2 per cent of cases) and discomfort they had suffered in the name of science).

Even though these were the preliminary findings of its investigation, the panel's conclusion was damning: 'This kind of error is a grave act that damages the foundation of science.' A statement issued that day from *Science* bluntly referred to 'substantial research misconduct'. With junior researchers sobbing behind him, an ashen-faced Hwang apologised but reiterated that his team was capable of the feat.

In light of these astonishing revelations, the panel said it would now also investigate Hwang's other landmark papers – which include another *Science* article in 2004 on the world's first cloned human embryos, and an August 2005 paper in the journal *Nature* on the first-ever cloned dog, called Snuppy (Seoul National University puppy), an Afghan hound. The journals were by then already conducting their own investigations. To establish if Snuppy was a clone, *Nature* had commissioned Elaine Ostrander, at the National Human Genome Research Institute, in Bethesda, Maryland, USA, to conduct DNA tests. *Science* had contacted the co-authors of its Hwang papers and asked MizMedi to launch an investigation.

During coffee breaks at laboratories around the planet, scientists gossiped about the scandal. How did fraudulent work end up being published? How could one keep fabricated data secret in such a large team? Perhaps Hwang's team cut corners and, confident it was capable of the feat, had assumed it could restore the lines that had been lost to contamination. Perhaps this affair exposed the weakness of Korea's science culture, where junior members who operate within a rigid hierarchy are not allowed to countenance failure.

By the end of December, any doubt about the last two of the eleven patient-specific cell lines in the 2005 paper had evaporated: 'We cannot find stem cells that have identical DNA fingerprint traces with patients and Hwang's team does not have scientific data to prove they did harvest patient-specific stem cells,' said a spokesman for the panel. All the samples presented for Hwang's paper, and even those Hwang said he had created after submitting the paper, were stem cells extracted from fertilised human eggs. Hwang had claimed his stem cells from cloned embryos had been switched. His accusations were met by counter claims and this aspect of the affair is still being investigated as I write.

At the start of January, *Science* announced that the 2005 Hwang paper had been retracted. The final verdict was delivered a few days later by the Seoul National University Investigation Committee: Hwang had indeed personally approved the donation of eggs by one of his graduate students and even accompanied her to MizMedi Hospital himself. A form asking consent for voluntary egg donation had been circulated around his laboratory and signed by technicians.

Hwang's embattled team's fraudulent experiments reached back even further than previously thought and encompassed the most seminal of his so-called successes: the first creation of stem cells from cloned human embryos, described in the 2004 *Science* paper. The first report of a stem cell line from a human cloned embryo resulted from one created by parthenogenesis, not nuclear transfer. The committee concluded that the DNA fingerprinting analyses and photographs of cells 'have also been fabricated'. The claims were, the panel said, 'scandalous'.

There was, however, some good news in the report. The panel concluded that Hwang's team did have the means to clone human blastocysts. And Snuppy looked genuine too, which was impressive given that the dog's reproductive cycle is nothing like as well understood as in the case of pigs, cattle and sheep. To investigate the world's first cloned dog, the panel had obtained samples from the egg donor, blood samples from Snuppy, from Tie, the dog that provided the cells for nuclear transfer, and from the surrogate mother and engaged three independent test centres for the analyses. To the relief of the journal *Nature*, the team concluded that Snuppy was indeed a clone of Tie and concluded that, 'when it comes to animal cloning, with the added consideration for the successful cloning of a dog, Korea seems to be internationally competitive'.

At least the depressing saga demonstrated that science – aided by investigative journalism – is relatively quick at weeding out fraud. Overall, the system of checks and balances in research works. As the panel pointed out, 'The young scientists who courageously pointed out the fallacy and precipitated the initiation of this investigation are our hope for the future.' But, like many other scientists who had visited Hwang and been impressed by his team, his science and his dedication, I felt badly let down. Although details of what actually happened are still emerging, the deception meant that embryonic stem cell research was years behind where scientists thought it was – in January 2006 it was all hype and no substance. In terms of creating stem cells from a cloned human embryo, we were back to square one. I have no doubt that this saga will be seen as a fiasco, one that can only be a setback in a field that has already triggered deep ambivalence and unease.

Therapeutic Cloning for Diabetes

While the Koreans appeared to be overtaking us, the first application made in the West to clone human embryos, part of an effort at Newcastle University led by Alison Murdoch and Miodrag Stojkovic, had been received by the UK's Human Fertilisation and Embryology Authority. By that time British stem cell science was enjoying a 'brain gain'. Stojkovic had already cloned cows, rabbits and mice, and arrived in Newcastle upon Tyne from the University of Munich in 2002, one of several stem cell researchers – including Roger Pedersen and Stephen Minger – who moved to Britain to take advantage of its more favourable regulatory climate and better funding. Stojkovic's team at the Institute of Human Genetics had already succeeded in deriving stem cells from non-cloned human embryos, and now wanted to produce the UK's first cloned embryo to develop treatments for diabetes.

Although the UK is a liberal regime for stem cell research, all such work had to be scrutinised. Five members of an HFEA committee studied the proposal, after inspecting the Newcastle lab and quizzing the team. Others asked questions too. Elliott Cannell, speaking on behalf of the pro-life group Comment on Reproductive Ethics, and himself a diabetic, said, 'One of the greatest obstacles for a starter is how does anybody imagine they will ever find sufficient human eggs to cure the 170 million diabetic sufferers worldwide. The HFEA has an absolute legal and moral duty to reject this cloning application.' I cannot imagine that any scientist engaged in stem cell research would dispute the idea that therapeutic cloning does not seem very practical as a mass treatment. But we are confident that it will show other ways to achieve the same ends, as well as creating type 1 diabetes cell lines that will accelerate the pace of discoveries that will in turn ease the suffering of diabetics. Even if it helps only a fraction of the 170 million suffers, it will be worth it.

The Newcastle project aimed to use some of the 1500 to 2000 eggs left over from fertility treatment annually at the Newcastle Fertility Centre, Centre for Life, with the full consent of the donors. As Stojkovic put it, 'I completely understand the ethical objections, but we are using eggs that are surplus to IVF treatment which failed to be fertilised. Instead of being thrown away they have been donated for research. The way I think of it is, why put something in the rubbish bin when it can be used in such a valuable way?'

One purpose of the new project was to improve the overall efficiency of therapeutic cloning. As Hwang found, it soon became apparent that fresh eggs were necessary to make nuclear transfer work. Another aim was to reveal new details of the disease process by comparing cloned diabetic cells with normal cells. Alison Murdoch, the director of the Newcastle Fertility Centre, voiced the hopes that propelled their research, using the example of a ten-year-old Newcastle child with diabetes. 'The ideal situation is that in ten years' time, when he is twenty, we could take a skin cell, reprogram it, make safe islet cells [which make insulin] and put them back into that boy so that he never needs an insulin injection again. That is why this work is so important.'

When, in August 2004, the Newcastle team was eventually granted its licence – the first in Europe to allow human embryos to be cloned – the image of therapeutic cloning as a potential cure-all suffered a blow when it emerged that the overall aim of the original application had been questioned by an HFEA committee. Although the HFEA was typically vague about what exactly was causing the bother, exaggerated claims for the possibilities of a treatment – discussed below – could have raised legal issues which the pro-life campaigners could have seized upon. As a consequence, the

Newcastle team dropped the overall aim of its project and focused on nuclear transfer alone. In April 2005 the HFEA eventually gave permission for the original aim of cloning cells from a type 1 diabetic.

I was pleased that the Newcastle team had been given the go-ahead for its work, but irritated because there was essentially no difference between what they had been allowed to do and what I and Christopher Shaw had asked to do years before: the HFEA had shifted from insisting that cloning be used for a specific treatment to allowing it to be used to lay the foundations for research into a serious disease. While I had contributed my fair share to delays, the HFEA's about-face was annoying. We could have started our programme of work on motor neuron disease as early as 2002. As it was, we were given permission in February 2005. Even that did not mark the start of our work, only the start of the effort to raise funds for the project. Christopher Shaw and I hoped to collaborate with Hwang's team, but the South Korean cloning scandal cast a long shadow over this work.

The kind of treatment envisaged by the team at Newcastle could have improved my father's life immeasurably if it had been available half a century ago. First, I would take a sample of his cells. They could be from the root of a hair or a swab of his cheek. I would transfer all the genetic material from his cell into a human egg whose own DNA had been removed. After the transfer, the egg with its adult nucleus would be stimulated with a shock of electricity to reprogram itself with my father's genes and start growing into an embryo/placenta. Then I would have derived his islet cells from the embryonic stem cells by means of the appropriate blend of growth factors. But there is one complication.

In many cell therapies it is an advantage to be able to grow tissue from a genetically identical embryo: that way, there is no risk of rejection. Thus, in my father's case the islet cells would have been compatible. However, diabetes of the kind that affected my father is caused by the destruction of the insulin-producing islets by the person's own immune system – which is why it is called an autoimmune disease. And we know from a study published by a team at the University of Minnesota in 1989 that replacing the islet cells with other genetically identical cells does not work as a treatment. When one of a pair of identical twins developed diabetes, islet tissue was transplanted from the healthy twin to the diabetic. Unfortunately the disease developed again. The immune response attacked the transplanted cells, and it is possible that cells from my father's cloned embryo would also have been attacked.

Initially, at least, it seems that there may be no advantage in having immunologically identical cells for treating diabetes. But there might be ways around this. One approach could be to genetically alter the stem cells in some way (either by genetic modification or by altering the genes that are used by the cells with growth conditions) so that the islet cells they make can be tolerated by the body. If he had received a transplant of modified versions of his own islets cells, my father's body could have manufactured its own insulin, sparing him all the awful complications of his diabetes, from amputations to blindness.

In many other cases the ability to grow cells and tissue that carry no risk of rejection would offer huge advantages. Heart muscle cells derived from embryos could be used to patch a damaged area of heart tissue. This has been done successfully in experimental animals. There would be a real advantage in using cells that are genetically identical to the patient's, for the patient would then not have to take immunosuppressive drugs. The same goes for any other muscle or skin. This kind of work is one of Dolly's greatest legacies.

There is one problem, however. Some people believe that this is tantamount to killing one person to save another. That complaint has dogged the field of human embryonic stem cell research every step of the way, reflecting the continuing debate over the moral status of the embryo. In response to the launch of the stem cell bank, one pro-life group called for the regulatory agency to do 'everything in its power to stop unnecessary destruction of these tiny human lives'. Its adherents fear that this kind of work will lead to the commodification and cannibalisation of embryos. They maintain that a fertilised egg, and everything that follows from it, is sacred. They oppose everything that I propose to do.

Chapter 7

Is a Blastocyst a Person?

What exactly do we mean when we use the word 'person'? What precisely are we referring to when we talk about a 'human being'? To some people, even the hollow ball of cells that forms a few days after a human egg is fertilised is a 'person', a 'tiny human being' who deserves the same rights as any other human – including, most fundamental of all, a right to life. The very thought of producing one by the process of cloning is repellent because it marks the manufacture of a 'new kind of human being'. Laboratory cloning is a 'violation of the natural order' and a 'repugnant manipulation', one that represents a 'trivialisation' of human life. Scientists like me who want to do this are portrayed by our opponents as having a 'lust for forbidden fruit'. We are, in effect, 'playing God' and exceeding our authority. We do not have the right to create, experiment on or destroy human embryos. My critics believe that it is not curiosity and a desire to do good, but hubris and 'hunger for power' that are driving me and my fellow scientists to drag us all into a grim Frankensteinian world.

Critics are horrified by the thought that a human egg can be fertilised in the laboratory, then grown for a few days for experiments. Or that healthy embryos can be selected for a pregnancy and unhealthy ones discarded. Or that an unfertilised egg can be used for cloning. Some even disapprove of using *in vitro* fertilisation to overcome infertility. This kind of vociferous opposition to human embryo research has already held back efforts to create new treatments based on stem cells. In the United States, under President George W. Bush, federal funding was withdrawn from studies with human embryos because of opposition from the Christian Right, though a vast amount of private- and-state funded work would continue to be done. In some European countries, such as Germany, France and Ireland, the creation of new lines of human embryonic stem cells was banned, creating problems for the European ideal of free exchange of ideas and people across borders. As I write, scientists are agonising over whether a German researcher who comes to work on human embryonic stem cells with me or other British teams could face legal action when she returns home. John Harris of the University of Manchester is one of a team of ethicists who are struggling to coordinate the ethical stance across Europe with the 'Eurostem' project.

At the heart of the furore lies a long and complicated dispute about the early human embryo. This is an argument about the status of the human blastocyst, the early embryo that has yet to implant itself in the wall of the uterus. The draft European ethical guidelines drawn up by Harris and his colleagues recognise this by, in effect, agreeing that the status of the blastocyst is the one thing on which everyone can disagree.

I am no moral philosopher, but I have also had to think hard about this issue, as I am sure most people have. But the questions that trouble the public are not always the ones that bother the specialists. The latter, be they scientists or moral philosophers, have jargon, terms and language of their own, making it difficult for members of the public to follow their debate. They may feel that certain issues are already settled. But that does not stop the same questions from being raised in the public debate, in appropriate language. I want to focus here on the concerns of the layperson, which have been voiced to me again and again at conferences and meetings.

The arguments about the critical milestones of a human life have been driven in great part by discoveries about events in the womb. During the first months of pregnancy, the mother sees her periods end and her breasts expand, she feels nauseous, and so on. In the days before medical diagnostics, the first obvious sign of the unborn child came with quickening – the first detectable movements of the foetus between seventeen and twenty-two weeks, described by some mothers as a fluttering feeling. Many societies past and present have regarded that stage as being the point at which the conceptus deserved full protection. Over the past century or two, medical science has given us many important new insights into the miracle of development. The German-Estonian Karl Ernst von Baer (1792–1876), who had previously discovered the ovum (egg), showed that the union with the sperm produced a pronounced change in its appearance. This provided a point of sharp discontinuity, a line in the shifting moral sands, and it was easier to regard this as marking the dawn of a life than any later stage in the development of the foetus.

Birth remains much more important to society than conception, however. This is the time of a baby's first breath, its first look at the world, its first cry. This is when the child is given a name and accepted as a member of the family. This is the event that is celebrated every year on birthdays. Yet even here, science provides new information that can trigger debate. New ways of taking pictures of the foetus in the womb, with sound waves, have stirred deep feelings in the hearts of many parents, feelings that their great-grandparents would have experienced only when witnessing the moment of birth. Ultrasound scans have shown that a foetus below the age of abortion appears to be able to do some of the things that babies can. Meanwhile, medical advances have continued to lower the age at which premature babies can be helped to survive, so that in one part of a hospital doctors can struggle to save a twenty-four-week-old foetus, while in another they can terminate it. In Britain these advances in understanding have quite rightly revived discussion of the morality of abortion at later stages of pregnancy.

Reproductive techniques can be equally contentious. Only the older ones among us will remember the furore over artificial insemination that took place fifty years ago. It arose first in the context of cattle breeding and became more relevant and heated when artificial insemination was then used to overcome human infertility. There was also intense debate about IVF after the birth of Louise Brown in 1978, now all but forgotten. IVF was viewed as crackpot science, and its pioneers as outsiders in the medical world who were interfering with an intimate process best left to nature. Like me, they were seen as taking the first step down a 'slippery slope' leading to eugenics, deformed babies and doctors playing God. It is reassuring to see how quickly society can develop an informed view so that a subject ceases to be contentious.

The current public debate about the significance of events at the beginning of life is reminiscent of earlier discussions about the end of life. When the first heart was transplanted from one person to another, in the 1960s, much was said about the circumstances in which it might be acceptable to remove an organ. Much of this debate focused on the effects of brain injury – 'ideal' donors would have a healthy body (and thus healthy organs) but a dead mind. The problem was how to tell when that hard-to-define property of consciousness had ebbed away. Again, it was new scientific observations that helped people to form a view. Most would now accept that we should use reactions to stimulation, measurements of flickers of electrical activity in the brain, and scanning techniques such as functional magnetic resonance imaging (which, as the name suggests, studies brain function) to determine the moment of brain death. The earliest stages of human life require a similar analysis.

What are the characteristics that we consider most profoundly human, and when do they emerge?

My proposal to carry out nuclear transfer with human eggs and cells has raised many questions. Various different positions have to be distinguished from one another and must be reconciled with attitudes to abortion, which are more focused on the status of the foetus, a much later stage of development. How do people form their views on these issues? Do they take their guidance from religious teachings, or do they draw upon some other ethical framework? Are these religious or moral issues? I know there are many religious views, passionately held, but for me the debate is a moral one.

We live in a secular society, and the question of research on embryos did not arise before the late twentieth century, so the status of the embryo in religious traditions can be deduced only indirectly from the Bible, the Koran and other religious texts. Over the years, religious convictions have evolved and changed, often in the light of scientific evidence. A layman might gain the impression that the Catholic Church has held a consistent line over the centuries on the central subject of human life, but in fact it was only as recently as 1869 that Pope Pius IX decreed that human life at conception was sacred, although the events during fertilisation would not be well understood for many decades to come.

Even today, some claims made by religious groups are clouded by misconceptions about the various stages of human development. That is not to say, however, that I am hostile to people who hold them. (If I were, I would not have married my wife, who is now a church elder.) Nor do I dismiss religious views. Indeed, the Christian tradition, for much of its history, distinguished between the moral status of the unformed embryo and the formed embryo, a distinction which I feel still has great validity. Since we know that a morally important person may eventually emerge (a substantial proportion of early embryos – some estimates put it as high as 50 per cent – are naturally lost before term), I believe that it is appropriate to accord a gradually increasing moral status to the embryo or foetus, tempered by the recognition that if there is some 'threshold of personhood', there will always be argument about where it lies.

In considering the proposals for my embryo research, we must ask a number of questions. Are all stages of human development equal, or does moral status develop from the moment of conception, through the embryo and foetal stages, to the birth of a child? Can the early human embryo suffer? Is the blastocyst conscious? Can it be called a person? The theologian and philosopher Michael Banner, of Edinburgh University, expresses the essence of the debate in the question 'Is an embryo a thou or a that?' If we accept that an early embryo is a 'that', and so lacks feeling and consciousness, the issue is then whether we owe the early embryo respect of such a kind that we ought not to produce and destroy it in the name of science.

I would like you to take stock of what you think about the moral status of the early embryo. Imagine how you would behave in the following scenario. You are visiting an IVF laboratory in the company of a young child. A laboratory assistant enters, carrying a Petri dish extremely carefully so as not to upset its delicate contents. The girl is fascinated to see, through a microscope, twelve early embryos in the dish, each consisting of eight cells. Two embryos are transferred to a patient, and the other ten are to be frozen and stored in a Dewar flask for possible later use. Suddenly, before anything can happen, the fire alarm goes off. Smoke billows in from the corridor, and you have moments to squeeze through a small window before the lab fills with fumes. What would you do – carry the child or the culture dish containing the ten embryos? (I should emphasise that it is not possible to move both, because the dish must be carried with great care.)

Some believe that the early embryo must be regarded as a human being in the fullest sense from the moment of fertilisation, and accorded the same respect as a baby. Thus the destruction of the embryo should be banned. Presumably, those who hold this view may be tempted to take the Petri dish, for it contains ten lives. Others regard the early embryo as little more than a collection of undifferentiated cells, which deserves no more attention than any other isolated human cell or tissue. And there is the middle ground where the embryo deserves some degree of respect because of its potential to develop into a human being in certain circumstances. If this is accepted, then we have another problem: what does it mean to show an embryo respect? Perhaps we can shed some light on these issues by first considering what it is that makes humans special.

Why Are Humans Special?

Some objections to the use of human embryos rest on the notion that there is something 'special' when it comes to human life. I agree that we are special, but feel that there is more to this than the basic attribute of being alive. From the perspective of a one-celled creature, to say that a human blastocyst is merely living is to do it a great injustice. Like a bacterium, each cell in a blastocyst reproduces and 'eats'. But it does something altogether more fascinating, something that deserves profound respect. When its cells split, they do not spawn new organisms but stay linked with the old ones to form a complex, interacting community. One of the greatest mysteries of all is how parental genes in a fertilised egg guide it within the womb to develop into the complex and finely integrated community that is you or me, consisting of two hundred or so different tribes of cells.

We know the process in rough outline. First, the fertilised egg splits to become two cells. Then they both split again to become four. They split once more to become eight, then sixteen, and so on. Repeat this process of cleaving in two around forty-seven times in all, and you end up with an intertwined and interacting super-society of ten thousand trillion cells. The actual number in an adult person is significantly fewer than this because nature kills off many of the cells by a process called apoptosis, making it hard to estimate the final tally. Through apoptosis, the body removes webbing between fingers, knits together the palate of the mouth, carves out toes, shapes organs, removes defective cells, and so on. The result is a complicated, cooperating three-dimensional whole formed of (fewer than) ten thousand trillion cells. This cellular community is what we call a person. But it has to be stressed that, although this person began as a blastocyst, a person does not count as a blastocyst, any more than a blastocyst counts as a person.

At the level of the cell, are we any more special than other creatures? Certainly more so than bacteria, which have a different design (in technical terms, they lack a cell nucleus). But comparisons of our genetic code with those of other multicellular organisms reveal that many of the genes we carry are dedicated to making the vast collection of ten thousand trillion cells work together as a coherent whole in all these different tribes of brain, hair, skin or whatever. The more important the role of particular genes in running the basic biochemical machinery of life, the more likely we are to share them with other creatures. That is why we share about half our genes with a banana, for example. As a result, at the genetic level we have a surprising amount in common with apes, with yeast, and with pretty much any other life form you can think of.

At this fundamental chemical level, you, I and the rest of *Homo sapiens* are different but not particularly special given what we know about human evolution and ancestry, let alone the astonishing similarities between our species and others. A reductionist alien scientist gazing only at the workings of human, yeast and banana cells would decide that the building blocks of a human are more similar to than different from the building blocks of apes, mice and a host of other creatures.

However, when it comes to the phenotype of a creature – the outward physical appearance and constitution of a creature as influenced by DNA in the nucleus, the way that creature looks and what it can do – *Homo sapiens* is distinct. Now we have moved from the basic properties of a cell to what are called the emergent properties of trillions of cells. We may not be able to outrun a cheetah, fly like a bird or bask in superheated waters near hydrothermal vents, but we are quite distinct and special in what we can do. And there is one important respect in which I do feel that humans are particularly special, not in the sense of any attribute that they alone possess, but in something of which they show an extreme manifestation.

I suspect that all animals, from the lowliest insects to the most sophisticated apes, experience varying degrees of consciousness and self-awareness. Precisely where the line should be drawn between a creature that simply reacts and one that can really think is a matter of intense debate, but there is no argument at all when it comes to the view that human consciousness is a special property of the cosmic numbers of nerve cells packed between our ears. From their activity emerges the property of self-consciousness, the ability to reflect and to reason. This is the crucial point to which I shall return when I come to consider whether the embryo should have the same rights as an adult.

The Birth of Genetic Identity

In our quest to weigh the moral status of the early embryo, let us start at the very beginning of when an individual is created. You might expect that when sperm fuses with egg, that is the very instant when a new individual is born. (By this I mean an individual with a distinct genetic identity, compared with that of the egg and sperm.) Yet studies reveal that the significance of the point at which sperm docks with egg is much less clear-cut. The genetic cargoes of egg and sperm initially lie dormant and separate, cloistered in different structures called pronuclei. These are residual egg and sperm, if you like, and the DNA of the new individual has not yet assumed control of development. In the early days of IVF, in the 1960s, I can remember a debate being triggered by a photograph published by one of the pioneers of IVF, Bob Edwards. The image clearly showed the two pronuclei, and other scientists present began to argue about whether it really counted as an embryo, whether fertilisation had occurred. Even in that scientific context there was debate about exactly what we mean by describing something as 'an embryo'.

Only when the pronuclei have come together and the fertilised human egg has divided three times do the genes of the embryo take control of development. Only then can we talk about the genetic recipe that will define a person. Only then are the instructions in that recipe acted on; before that point, instructions from the mother and father – in the form of RNA and protein – help to direct the first stages of life. Only after this stage does an individual's DNA take over. One could argue that only at this very moment is an individual present in the most basic sense, that of a distinct genetic recipe.

The blastocyst that sits silently at the heart of this debate is a far cry from the popular image of the embryo, of a little foetus with limbs and heart, brain and other organs. In reality, this stage forms after around seven weeks of development, when all the individual organs become recognisable. The blastocyst stage of life that forms a week or so after fertilisation is nothing like this: it consists of a ball of cells stuck together like luminous tapioca, or frog spawn. It is just visible to the naked eye, but smaller than a grain of sand. When I look at a sheep blastocyst under the microscope, as I have done a thousand times, I do not think of it as a sheep. Nor would I think of babies when I gaze at a human blastocyst, though I accept that some couples undergoing IVF treatment do become attached to their early embryos.

Each human blastocyst consists of a hundred or more cells. The much sought-after embryonic stem cells are among the twenty-five or so in the inner cell mass, the clutch that are destined to become the foetus and some of the placenta that lie on one side of the hollow sphere of cells (the trophoblast) that becomes the rest of the placenta. Unlike adult stem cells, they have the ability to turn into any part of an embryo and thus make any cell type in the body. Embryonic stem cells, like people, respond to their culture, surroundings and upbringing. But the conditions in which embryonic stem cells exist in a laboratory can never fully mirror those experienced by cells in an embryo developing within the womb. Although biologists have been studying and using mouse embryonic stem cells for twenty years now, we do not understand the details of why these cells, derived from the inner cells of a blastocyst, are so extraordinarily versatile or how they have the ability to turn into any other type of cell.

Once separated from their sister cells, these embryonic stem cells lack the crucial cues and signals to instruct them to develop into a foetus. When grown in a suspended culture of proteins and hormones, for example, they form what is called an embryoid body, a spherical structure consisting of a clump of cells that differentiates in a disordered way, so it does not count as an embryo and cannot develop into a foetus. An embryo needs the environment of the womb if it is to develop at all. In a Petri dish it has no potential to do this.

An early embryo is unlike you or me or any other individual person in other respects. One blastocyst can split into two, as happens naturally to produce twins. Conversely, if there are two embryos, they can combine into one early in pregnancy, as happens naturally in the womb, albeit rarely, to create a chimera. Or a blastocyst can – in theory at least – split into two and recombine again. Did 'life' in such a case begin as an individual, become two individuals, and then turn into a singleton again? There are other peculiarities. What many do not fully appreciate is that three-quarters of the cells in the blastocyst, around seventy-five, are destined to make the first contact with the lining of the womb and contribute to the placenta. As highlighted by the people who use the placenta to fertilise plants, consume it as a postpartum snack (browse the Internet and you will find plenty of recipes) or use it as a medicine, society does not accord the progeny of these cells the same respect as the embryo proper.

None of the above points mean, of course, that an embryo is not an individual. Nor do they suggest that an embryo does not deserve respect. I merely point out the messy details to show that when we do get the full picture of the first stages in a human life, the identity of the early embryo does not map neatly onto the identity of a person.

A Blastocyst Is Not a Person

Although a human embryo has only just begun to develop, it is deserving of our respect because in the right circumstances it has the potential to become a person. However, just because a blastocyst has this potential that does not mean that it *is* a person, just as a young girl who wants to study medicine is not a qualified doctor. As John Harris argued in *The Value of Life*, we are all potentially dead, but that does not mean that we should be treated as if we are dead.

Nor does a fertilised egg created by scientists in itself have any potential to become a person without human intervention. The early embryo sits not in a womb but in a Petri dish and is difficult to grow in laboratory conditions for more than five or six days.

Until a naturally fertilised egg attaches itself to the wall of the uterus, where it starts a highly complex chemical dialogue with the mother, there is no pregnancy. A blastocyst sitting in a Petri dish goes through no further stages of embryonic development. It has the potential to grow into a foetus or a baby, but only with human assistance.

The main reason why I do not regard a blastocyst as a person is that it has no mental life. In an adult person there are around a hundred billion synapses, the connections between nerve cells. Each cell has around a thousand connections with other cells, so there are about a hundred trillion connections that store information by a process called plasticity (when one nerve cell 'talks' to another, it can modify how efficiently communication occurs, and this is believed to be critical for brain function). A working human brain is thought to make an average of a million connections every second. The critical issue is when this capacity to think first appears.

In effect, we are groping for the idea of what some call 'brain birth', the mirror image of 'brain death'. Exactly when this quality emerges is the focus of much debate. This was highlighted by an expert panel that discussed proposals to grow human neural stem cells in the brains of developing non-human primates to provide better models of Parkinson's and other neurological diseases. The panel, which included Davor Solter and John Gearhart, concluded that the implants could affect cognition in such a way that the resulting animal would have a moral status closer to that of humans. Importantly, however, there was no consensus on the moral significance of changes in cognitive abilities, even if we could detect them.

John Harris argues that a 'person' is 'a creature capable of valuing its own existence'. This, he says, makes plausible an explanation of the nature of the wrong done to such a being when it is deprived of existence. Over a lifetime, an individual will gradually move from being a potential person – a 'pre-person' – into an actual person when she becomes capable of valuing her own existence. 'And if, eventually, she permanently loses this capacity prior to death, she will have ceased to be a person.' This is all very logical and coherent, but I suspect that most people would regard an elderly relative who had lost their mind to Alzheimer's as still being a person. For many, the same would go for individuals who are 'brain-dead' – anencephalic infants, who lack part or all of the brain, or individuals in a persistent vegetative state. And by the same token I would have an equally conservative view of personhood at the start of human life. Where, precisely, I cannot say. But of this I am certain: the clutch of one or two hundred cells that makes up a blastocyst is *not* a person.

No feelings or sentience will stir within the blastocyst without a functioning nervous system. Although stem cells are removed from an embryo around six or seven days after fertilisation, it is only at around fourteen days, when it consists of several thousand cells and is still no bigger than the full stop at the end of this sentence, that an embryo develops the so-called primitive streak, the first glimmer of the machinery of thought. This is the precursor to the spinal cord and backbone and central nervous system that will develop later. It is also beyond the point where twinning can occur.

For me this is an important milestone in development. Now, at last, we have reached a point where the embryo is committed to creating a single human being. Crucially, this point can be defined objectively, in a way that, in the words of the great theological thinker Dietrich Bonhoeffer, allows me to take up the 'sharp swords' of simplicity and wisdom.

This is why I feel I can justify the use of blastocysts in my research. In turn, this raises other questions familiar to anyone who does research on living creatures. Is a blastocyst aware? Can a blastocyst feel pain? Until nerve connections form between two crucial areas of the developing brain, the cortex and the thalamus, sensations of pain cannot be experienced. This formation takes place much later in a pregnancy, at around the twenty-sixth week. There is no pain, no suffering, and so no cruelty to the blastocyst is possible.

Respect the Embryo

In the light of the above discussion, I understand and back the current UK regulations that prohibit the use in research of embryos more than fourteen days old. The appearance of the primitive streak marks a very conservative starting point for a person, one that gives the benefit of the doubt to those who feel uneasy about embryo research. I believe that the use of blastocysts before this point in development can and should be justified in carefully regulated research and in certain circumstances. That is not to say, however, that an embryo younger than two weeks is not deserving of some respect.

Although the respect shown an early embryo falls far short of total protection, British regulations do reveal this respect in ways that I fully support: early embryos should be used in research only if there is no alternative, and only with consent, and meticulous records must be kept to ensure that every embryo can be accounted for. One should expect any research carried out on an embryo to be of the highest possible quality, and justifiable. As part of the respect I accord the embryo, I share with many others the view that we should do our best to use and investigate alternatives, such as adult stem cells or parthenotes.

Despite the deliberations on this issue by the Roslin group, politicians and many others, I know I will be criticised. At one extreme are those who will still maintain that my research on human blastocysts is tantamount to killing people. At the other are those who would find my regard for the early embryo woolly-headed. About my respect for the early embryo, John Harris asked,

'Is that like saying that I respect the pig in the bacon sandwich and am doing my best to find something I like to eat as much as bacon sandwiches?'

We should be realistic about whether it will ever be possible to reduce complex ethical and moral issues to pure logic, however. The work of the Kurt Gödel and others showed that it is not even possible to deduce all mathematics from the axioms of logic, leading one mathematician to joke that, if we define a religion as a system of thought that contains unprovable statements, and thus an element of faith, then 'Gödel has taught us that not only is mathematics a religion but it is the only religion able to prove itself to be one.' Even scientists who reject religion must have some articles of faith.

Doubts, but Not About the Blastocyst

Despite the best efforts of philosophers, and despite my little contribution to the debate, science as it stands cannot come up with a definitive statement on the moral status of the early embryo. One huge hurdle is the absence of general agreement on what we mean by life. The eminent Austrian physicist Erwin Schrödinger suggested in his book *What Is Life?* that a fundamental property was the tendency of life to produce increased order and seemingly unlikely arrangements of things, whereas the second law of thermodynamics suggested that, left to themselves, things tend to end up in the most probable state, a disordered mess. Another idea is that life is intricately associated with evolution: life is a self-sustained chemical system capable of undergoing Darwinian evolution. Biologists can list a whole set of features possessed by nearly all living things. As well as the ability to reproduce, these include the existence of genetic information, complexity, organisation, and so on.

But exceptions can always be found. For example, the ability to reproduce is not possessed by every object we might expect to call 'living': sterile men, postmenopausal women, mules and viruses are all incapable of self-reproduction. A sneeze demonstrates how a common cold virus can persuade cells in the human respiratory tract to reproduce and spread it, even though it is little more than a piece of genetic program wrapped in protein. Non-living things also show some 'vital' signs. Crystals, for instance, are capable of self-reproduction during growth. And so forth.

The same confusion surrounds the critical issue of consciousness. We can all agree that the human brain is unmatched in its ability to think, to communicate and to reason. Most striking of all, it has a unique awareness of its identity and of its place in space and time. But consciousness is notoriously difficult to define or locate. Many cultures and religious traditions place the seat of consciousness in a soul that is separate from the body. Conversely, many scientists – I am among them – and philosophers consider consciousness to be intimately linked to the neural functioning of the brain. If we could define the moment when a developing embryo becomes conscious, and quantify this consciousness, that would lay down a crucial marker for the debate. I remain doubtful that this will ever be possible.

Some would argue that I cannot therefore prove, in the strongest sense, that a blastocyst is not a person, and that the early embryo should be given the benefit of the doubt; that even though it is probably not a person, it should still be treated as such and accorded the full rights of a baby. I would accept that this infinitesimal doubt about the status of the early embryo would indeed be enough to halt research. Once again, though, I have to say that there is no doubt in my mind that a blastocyst does not possess consciousness. Although the arguments in support of embryo experimentation fall short of the rigour of a mathematical proof, there are also morally weighty reasons to continue. I respect the early embryo and have certain duties to it, but I passionately believe that these are overridden by the duty we have to people.

The Rights of a Blastocyst's Parents

One social change in my lifetime that I applaud is the considerable lengths to which we now go to provide equality of opportunity. However, this is very different from saying that every embryo deserves the chance of life, as some argue, or suggesting that potential parents should not exploit modern medicine to prevent the birth of children with inherited disease.

Since the embryo is unable to make judgements, it depends on the 'parents' to act on its behalf, as indeed they would if it continued to develop into a child. The debate about reproductive technology often neglects to note that embryo research can continue only with the cooperation of the people who give their approval for their embryos to be used. Those people have rights, too. Indeed, even if reproductive cloning were allowed today in our democratic society (and I hope it never is), and even if all the scientists wanting to do it were the Nazi eugenicists that the critics fear, it could happen only with the approval of two or three people: the individual who is to be cloned, the woman who donates her eggs and the woman (perhaps the egg donor) who is to become pregnant. Feminist groups claim that women's bodies are being exploited in an unacceptable manner in these new reproductive technologies. The point is, however, that people do have the right to choose, to say no if they do not want their embryos to be used this way.

Some parents have confronted these decisions already, and as we learn more about human genetic diseases, more of them will. In many cases the embryo will have a genome that makes it vulnerable to an inherited disease, or even condemns it to a short life. Even now, IVF embryos can be screened when parents know that an awful disease stalks their family, so that only healthy embryos are implanted and the rest discarded. So-called pre-implantation genetic diagnosis, PGD, can deal with disorders with specific gene culprits, including cystic fibrosis, sickle-cell anaemia, muscular dystrophy, Tay–Sachs disease and fragile X syndrome. Newer techniques have enabled PGD to detect, for example, chromosomal aberrations that cause Down's syndrome and certain blood cell disorders.

The possibilities are expanding, and the use of these tests is a matter for the parents to decide: the impact of a test for a mutated gene, whether positive or negative, can have consequences not just for the embryo but for the family at large. And some of these consequences are not always obvious. Think, for example, of the middle-aged woman who finds out from her identical twin sister, who is undergoing PGD, that she does not carry a gene for the devastating degenerative disorder that killed their mother: she may well have mixed feelings if she has spent her life avoiding relationships with men and the pressure to have children.

What of surplus embryos that are produced during fertility treatment that are not known to be vulnerable to inherited disease? Sometimes a couple will be fortunate to have many healthy embryos produced, with some to spare even after they have had children. These are likely to be frozen and stored so that the decision about what to do with them can be delayed, sometimes until years after their production. The couple may choose to donate them to other infertile couples who are not so fortunate, but this requires them to accept that someone else will bring up the children that develop from those embryos. Others may choose to donate spare embryos for research, as has been done to create embryonic stem cell lines. Still others may decide to have the embryos destroyed. If the couple cannot contemplate the idea either of a child of theirs being born in a different family or of embryos being used in research, this will be the only option available to them. In Britain the law requires that frozen embryos be stored for a maximum of five years before they are thawed and either used or destroyed. The couple would have been made aware of this deadline before fertility treatment began.

Many people take a pragmatic view. They may not support the production of embryos specifically for research or therapy, but accept that embryos that already exist should be used rather than destroyed. As a result of the astonishing rise of IVF over the past two decades, many blastocysts are now left over from fertility treatment. Of the 763,509 embryos created by IVF between 1991 and 1998 in Britain, for example, 184,000 were stored, 48,000 were used in research and 238,000 were destroyed. Pragmatists argue that it is better to make use of this precious resource – with the informed permission of the couples concerned – rather than throw it away. They maintain that this accords the blastocyst greater respect. Indeed, they say, using spare blastocysts in research to develop new treatments and deeper understanding of serious disease is more than morally defensible: it is a moral obligation.

What, then, of the production of embryos for the specific purpose of research or therapy? In the dozen years after the enactment of British legislation in 1990, only 118 embryos were created for experiments, compared with the hundreds of thousands used in IVF. And rightly so, because the specific creation of a blastocyst for research should be a last resort in exceptional situations. Once again, the circumstances vary. A healthy woman may choose to donate her eggs in the knowledge that they will be used not to produce a child but to create cloned embryos with an inherited disease that offer new opportunities to understand the illness, test drugs, and so on. One embryo provides endless opportunities for research, for once a stem cell line has been established it can be grown in the laboratory for years. Perhaps she will choose to do this because she has seen the suffering of a child, friend or colleague and understands the potential value of the research. Others will know from personal experience of the suffering that this research could help to end. My group is asking people with inherited forms of motor neuron disease to provide cells that we can use in nuclear transfer to recreate in the laboratory the cells that were present in their bodies early in life. In this way we hope to get the first glimpse of the fundamental cause of what is going wrong. The donors recognise that there is only a limited chance that the research will assist them, but they want to help future patients.

Whatever the motivation, it is crucial that donors are well informed. That information must come from an impartial source, someone who does not stand to benefit from the research or therapy. Nor should there be any question of financial reward for donors. The circumstances will often be difficult and charged with emotion. The responsibility of the counsellor, typically a nurse, is to describe what will be involved for the donor – the physical discomfort they will suffer and the risk of any long-term effect on their health. The donors must understand exactly what will happen in the laboratory and to what stage the embryo will be allowed to develop. Those people who find it unacceptable to produce or destroy embryos must not be put under any pressure to change their minds. There must be time for the potential donors to weigh all the information and to reconsider if they so wish. Cell donors have a basic right to say no.

The Yuck Factor

Fundamentalist opponents raise many other objections to signal their disgust with embryo manipulation. Some claim that therapeutic cloning and test-tube babies diminish our humanity. Hardly. The extraordinary inventiveness that made nuclear transfer and IVF possible is a defining human characteristic. Many critics charge that embryo research is an affront to human dignity, but this argument seems to founder on what we actually mean by human dignity. Human beings deserve respect – but then so do all living things with the ability to feel and think to some degree. Is it undignified to use one's ingenuity to overcome infertility, to treat disease, to reduce suffering? I do not think so. In fact, the opposite is true: surely cloning replacement nerve cells to repair a brain succumbing to dementia is more dignified than leaving an Alzheimer's patient to lose his memories, his friends, his home, and even control over his bowel functions?

Many critics believe that my research is 'unnatural'. This complaint raises various questions about what we mean by 'natural'. When it comes to our home planet, for example, which vintage of Mother Earth is the most 'natural' and thus the one to which we should be faithful? The Earth as it was a hundred years ago, or a thousand years ago? Or before bacteria enriched its atmosphere with oxygen? Or perhaps a very new Earth, one subjected to the hell of the great bombardment of four billion years ago, when our newborn planet was sweeping up the cosmic garbage and debris in its orbital neighbourhood? And what about us? If we humans are natural, are the tools we develop natural too? And if not, does that mean that the pandanus leaf tool of the New Caledonian crow is unnatural? And how about the 'fishing sticks' used by apes, and those rocks that Egyptian vultures drop on ostrich eggs?

Many things which are commonplace in nature seem, well, unnatural when subjected to scientific scrutiny. When you have a streaming cold, a scrap of genetic material wrapped up in protein (a virus) has invaded your respiratory tract, genetically altering your cells so they become virus factories. For many people, the thought of alien genes taking over their noses may sound fantastic. But it happens every day. They may also be perturbed to discover that their cells are powered by microbial technology – structures called mitochondria, which were once independent organisms but then, some two billion years ago, set up home in the cells of our ancestors, trading energy for a comfortable and dependable home. Those people also rely on kilograms of microbes to digest their food. Indeed, some of these microbes interact with intestinal cells, called Paneth cells, to promote the development of blood vessels in the intestinal lining following birth. It sounds incredible – like a B movie science-fiction plot or a tabloid headline – but, from princes to paupers, we are all powered by bugs.

Many people still hanker after a more natural world. I would sympathise with this argument if it came from a hunter-gatherer who has – and wants – no dealings with computers, cars, antibiotics, disinfectants, or any of the other 'unnatural' trappings of modern life which allow us to live longer and more comfortably than ever before. But in the developed world the assumption that 'the way nature intended it is best' is a fallacy. In such a state of nature – or at an earlier time in our history that more closely approximates it than that of humankind in late industrial society – there would be many more deaths in childbirth, many more people killed by infection and much more suffering. No one would suggest a return to such a state on the grounds that it is more 'natural'. No one mourns the passing of smallpox. No one looks back to the good old days before pain relief. When we are heating our homes, driving to work or taking paracetamol when we have a headache, we conveniently ignore this fallacy in our daily lives, but the very mention of cloning often triggers passionate cries that it should be rejected because 'nature knows best'.

I could even argue that cloning (though not the process of nuclear transfer used to create Dolly) is the most natural thing in the world. After all, in the very beginning was the clone. Four billion years ago, replicating chemical reactions stirred on the newborn planet Earth. The seeds of life were sown by these reactions, in which molecules reproduced themselves to make 'chemical clones'. They began to straddle the watershed that divides multiplying molecules, simple replicators and other types of dead chemistry from animate, breathing biochemistry. Something emerged that we would today classify as living – a self-replicating chemical reaction swaddled in a protective, filmy envelope. Within this first organism, billions of dumb and purpose-free molecules engaged in a complex choreography, enabling it to cleave. When one became two, the first clone was born.

Echoes of the chemistry that turned within that very first clone can still be found within our own cells. The cleaving when the cells in our bodies divide can be thought of as a kind of cloning. The duplication of the genes during cell division can also be considered as cloning. Each of our body cells is a clone of all the other cells and of the original fertilised egg. Mitochondria, the chemical power packs that drive our cells, clone themselves to help drive on the next generation. We owe our most ancient origins and our continued existence to cloning. Then, of course, there is the 'clone' that arises from a technique called nuclear transfer, such as Dolly. This is a highly inefficient method for making a genetic copy of a creature, a second-rate imitation of the process that bacteria have been carrying out effortlessly for billions of years.

Of all the cries of 'yuck', one of the loudest comes from those who worry that if cloning embryos for therapy is allowed, we will inevitably end up with cloned babies. Insofar as using a six-day-old embryo in research places us on a slippery slope, there is nothing special in that. Our heels are skidding on a steep and slippery slope in our treatment of animals, in our legislation on abortion, and in that which governs sexual behaviour. It is the job of democracy to decide just how far down the slope we should be allowed to slide. I cannot deny that research on therapeutic cloning will inevitably help to hone techniques that can be used in reproductive cloning. But that is not to say that reproductive cloning will happen. It is banned in Britain, whereas therapeutic cloning is allowed, albeit under strict regulation. No matter how slippery that slope, it is easy to draw a well-defined line on it in this particular case. Society can have the good and reject the bad.

Power, Responsibility and Procreation

The basic fears stirred by the opportunities that scientific research creates are not new. The city where I live once staged protests at the introduction of anaesthetics during childbirth. Professor James Simpson was born in 1811 to the family of a poor Edinburgh baker, the seventh son of eight children. He was very intelligent and deeply compassionate. Appointed professor of medicine and mid-wifery at the remarkably young age of twenty-nine, he introduced the use of ether and chloroform in obstetrics. But he faced opposition.

In the early days, anaesthesia could be hazardous. To understand-able fears about safety were added religious objections. Pain was thought to be necessary for the normal progress of labour, while churchmen (and some physicians) held that anaesthesia, in abolish-ing the pains that constituted God's punishment on womankind for tempting Adam's fall, was sinful because it circumvented the chas-tisement inflicted by a Higher Power. Local clerics, who of course were all men, argued that women were intended to suffer in child-birth. But some religious leaders were supportive, and Simpson gradually won the argument, especially after Queen Victoria's use of chloroform for the birth of her eighth child, Prince Leopold, in 1853. The local clinic still bears Simpson's name, and it is said that when he died thirty thousand mourners lined the streets of Edinburgh. A plaque in Westminster Abbey reads, 'To whose genius and benevolence the world owes the blessings derived from the use of chloroform for the relief of suffering.'

Similar debates have raged about the application of knowledge ever since our ancestors first learned how to kindle fire. To ban all cloning – both therapeutic and reproductive – is akin to a Stone Age tribe banning any use of fire because you can use it to burn down your neighbour's hut, even though it also keeps you warm and cooks your food. The understanding of how an influenza virus killed up to fifty million people in 1918 can just as easily be abused to make a biowarfare agent as it can be used to create vaccines and drugs to halt a future pandemic. And the biology that will let people clone children, which may have unfortunate consequences for them (if they survive childbirth), will also help us to treat some very nasty diseases.

Every time I hear calls for blanket curbs on genetic technology, I think of the heart patient who died waiting for a transplant as animal rights protestors championed the rights of xenograft pigs, or the mother forced to watch her son suffer because fundamentalists took legal action to prevent her from having what they call a 'designer baby' (a technique that relies on embryo selection, not genetic modification, as this pejorative label suggests). Abandoning a particularly 'dangerous' technology wholesale can kill, maim and hurt future generations by preventing that technology from doing any good at all. Society has to weigh the opportunities to help and make sure that it does not miss important new opportunities because of fear of new knowledge. We should be prepared to change our views and judgements in the light of new discoveries.

Each advance in reproductive science sends shock waves through society. Perhaps the best-known example is IVF, where an egg is fertilised in the laboratory and the resulting embryo implanted in the mother. Indeed, without this research we would not be arguing about the status of the embryo. Apart from bringing hope to millions of childless couples, IVF has provided access to the hidden world of the human egg and the embryo.

If we have learned anything from Louise Brown, the world's first IVF baby, it is that the revulsion that greets cloning is likely to give way to the acceptance of some, but not necessarily all, of its uses when the public becomes aware of the benefits, such as the ability to grow replacement cells to repair a damaged body. Many of the major advances in biology have evoked that same knee-jerk outpouring of horror. The first deliberate transfer of genes between microbes and the first heart transplant are among them. These are now everyday procedures. I would like to think that the same will go for the use of nuclear transfer to derive cells or to correct human genetic disease. As I shall attempt to explain later, the treatment of genetic disease is one application of cloning to produce a child that may one day offer real benefits to society.

Chapter 8

Why We Should Not Clone Babies

When Lesley Brown held her daughter Louise for the first time, she was speechless. The expression on her face was absolutely extraordinary, remarked one of the doctors present at the birth, in Oldham and District General Hospital. After about a minute she turned to them and said, 'Thank you for my baby.' John Brown refused to hold his little daughter at first – he was shaking too much, on what he now calls the most beautiful day of his life. For scientists, too, 25 July 1978 was historic. On that day was born a healthy baby girl who was conceived in a glass Petri dish – the world's first test-tube baby. Since Louise, more than two million babies have looked up at parents who would otherwise have failed to conceive children.

Louise Brown's birth helped to launch a revolution in reproductive technologies and medicine that is still changing the world, challenging moral values and stirring religious debate. As one of the pioneers of *in vitro* fertilisation, Robert Edwards, said of their impact, 'They change an attitude. They change a way of doing things. They change an ethic. They change the way a nation looks at itself.' Louise's story remains extraordinary for the insights it gives us into the relationship between society and reproductive science, when the despair of childless couples – who constitute between 9 and 15 per cent of all married couples – meets the ambition and scientific curiosity of doctors. Her story is also highly relevant to the debate over cloning. Before she was conceived, human reproduction was regarded as almost sacred. For Edwards, it seemed to be a matter 'for theologians, rather than gynaecologists and scientists'. Despite the misgivings, IVF has become widely accepted because human misery is the driving force, as is the case with therapeutic cloning.

But will that also be the case with reproductive cloning, the 'copying' of people? While millions will be disgusted by the idea, there will be a few desperate people for whom cloning will raise hopes – all false. There will be some, like the biologist Richard Dawkins, who have genuine curiosity about cloning themselves ('I find it a personally riveting thought that I could watch a small copy of myself, fifty years younger'). There will be a small group of doctors hoping to exploit these desperate people to make a living, perhaps even a fortune, and secure their place in history. Others will talk about the need to be pragmatic and about circumstances in which it would be justified to clone a baby. But, as Bertrand Russell once said, 'Pragmatism is like that warm bath that heats up so imperceptibly that you don't know when to scream.' Many countries will have legislation in place to ensure that we never reach the boiling point of reproductive cloning. However, laws forbid; they do not always prevent.

I do not believe that the moral of the story of IVF is that anything in reproductive science goes, including the cloning of babies. As I argued earlier, in discussing Stephen Levick's work, there are good reasons not to clone. People have already cloned dead pets in the mistaken belief that the resulting genetic twin will be a kind of resurrection of their lost companion. Everything from the coat colour to the behaviour of these clones has underlined just how false this is. Because there is no such thing as genetic determinism, a clone could never replace a lost child. Levick has also pointed to the identity confusion that a child would go through on discovering that she was her mother's identical twin, or perhaps even her grandmother's. Others have warned of discrimination against clones, or 'clonism'.

Crucially, there was a real point to IVF, which had been driven by efforts to alleviate a common cause of heartbreak. Louise's mother and father, Lesley and John, were a couple who could not have children. At that time, one in eight couples could not conceive. The technique has since given millions of people hope. According to the British Fertility Society, nearly 2 per cent of all babies born in the UK during 2005 were conceived thanks to assisted reproductive technologies.

Some advocates of reproductive cloning have claimed that infertile men and women could now resort to nuclear transfer to have a child, or that homosexual couples could produce a child from a chimera blended from two cloned embryos. But I find it hard to imagine a scenario where the huge risks of cloning – outlined in detail in this chapter – can be justified, given the range of alternatives on offer: adoption, surrogacy and a huge range of infertility treatments.

If one accepts that the applications of reproductive science are driven by a utilitarian outlook, where there are tangible benefits to society and proportionately low risks, then there is no future in reproductive cloning. However, I can envisage one circumstance, outlined in the next chapter, where I think that birth achieved by nuclear transfer could be justified. In that case, however, the baby would not be a clone of a person.

After Louise

Maverick scientists often claim that the work on Dolly and other animals has laid the groundwork for human cloning. There are two things to say about this. First, as already discussed and well described by Stephen Levick, there would be unacceptable social consequences for a cloned child who was a genetically identical twin of someone else. Second, cloning research has raised serious concerns about the health of both the birth mothers and the clones themselves, given the low efficiency and high risks of nuclear transfer. Although nuclear transfer is becoming safer and more efficient, the experience so far gained by cloning animals shows that the possibility for harm is much greater than with IVF, so much so that the benefits would have to be much more obvious in order for those risks to be justified.

To make Dolly we started with 277 reprogrammed eggs. Only twenty-nine made it to the stage where they could be implanted, into thirteen surrogate mothers. Of those mothers, only one became pregnant, carrying Dolly. Some claim that this success rate of one in twenty-nine is not bad. They argue that reproductive cloning could even look attractive when it is remembered that these embryos have been screened for chromosomal abnormalities, a problem which affects around half of the eggs of women aged under forty and higher percentages in older women.

But no reasonable and rational person would really want to clone a child given the dismal track record of my team and others worldwide. To repeat the Dolly experience in humans would mean obtaining around three hundred eggs, which are already in short supply, and then persuading twenty-nine women to agree to having an embryo implanted. Of those, twenty-eight would go through the hurt and emotional turmoil of failed pregnancy, miscarriage and a deformed foetus in order that one embryo will 'take' to produce a child. To have only one woman act as the clone's surrogate makes the whole enterprise even more horrendous. To me, this exercise in human misery and suffering seems inconceivable in a free society. And given the differences between humans and sheep, the suffering that would be caused just to create one human clone could well be much greater. Even a sinister billionaire in an unethical society would find it tough to indulge a narcissistic cloning fantasy. To clone humans is criminally irresponsible.

Will Human Clones Have Big Belly Buttons?

Since the birth of Dolly there has been enough experience of animal cloning to build up a disturbing and detailed picture of the physical dangers of human cloning by current techniques, above and beyond the psychological dangers. Independent groups have now published comprehensive reports on cloning cows which show that safe reproductive cloning remains a distant prospect. One key research programme in cloning has been led by Jean-Paul Renard at the Institut National de la Recherche Agronomique (INRA) at Jouy-en-Josas, near Paris. Renard was the first to clone a rat (notably Ralph the rat pup) and the first to clone a rabbit. His team also provided independent confirmation that cloning mice is possible, and that cloning can be done by means of differentiated cells – the latter revealed when Marguerite the calf was unveiled at the Paris Agricultural Show ('Vive le clone!' as *New Scientist* put it in 1998).

Renard's team has worked diligently to compare the pregnancies that occurred in around three hundred cows after the transfer of IVF embryos with those produced after the transfer of cloned embryos derived from three sources of donor cells: adult cells, foetal cells and cells from embryos at the thirty-two-cell (morula) stage. As one would expect from the chequered history of cloning, the more differentiated the donated cell, the less efficient nuclear transfer becomes. So when it came to cloning an adult animal, only nine calves developed from 133 blastocysts, an efficiency of just under 7 per cent. Many more embryos would have been produced, but failed to develop to the blastocyst stage.

In similar work, David Wells and Björn Oback from AgResearch in New Zealand have found that from 988 somatic cell-cloned embryos transferred into recipient cows between 1997 and 2003, 133 calves were delivered at term, representing a 13 per cent survival rate. Only 89 of those calves survived until weaning at three months of age, reducing the overall survival rate to 9 per cent. This means that clones still have less than a third of the success rate for bovine IVF embryos, where about 30 per cent of transferred embryos develop into healthy calves at weaning.

There is an unpleasant statistic of this low efficiency of bovine cloning that any budding human cloner should bear in mind. In most cloning experiments, fewer than half the embryos survive for even a week, by which time they should have reached the blastocyst stage. By contrast, almost all eggs fertilised after natural mating reach this stage of development. Clone pregnancies can fail at any time during gestation, while IVF pregnancies tend to fail early, if they fail at all: in IVF almost all embryo losses take place by day 35, which is early in the 280-day pregnancy of a cow.

Because embryos produced by nuclear transfer continue to be lost throughout gestation, it would not be uncommon for more than 10 per cent of cloned embryos that have developed success-fully for the first third of pregnancy to die before the end of pregnancy (and twice as many for cloning with adult cells). This is true in all species that have been studied, regardless of labora-tory, choice of donor cell, method of nuclear transfer or anything else you can think of. The toll of miscarriage is high. Overall, the proportion of embryos born as calves in the two studies in France and New Zealand was fairly similar, at 7 and 13 per cent, respec-tively.

The abnormalities that commonly arise during a clone preg-nancy include overweight foetus or placenta, fluid accumulation within the placenta (particularly in the allantois, the sac-like mem-brane that develops into the umbilical cord and part of the placenta), inadequate attempts by the mother to give birth, an inability of the newborn animal to begin to breathe (it is unable to push blood through blood vessels that are many times bigger than normal). Less common abnormalities include failure of the immune system to develop and abnormal development of organs such as liver, kidney and brain. I am sure there are more horrors because, unsurprisingly, laboratories have been reluctant to describe all the abnormalities they have encountered.

The problems of cloning continue after birth: the viability of cloned animals is significantly lower than for animals conceived naturally. In our first studies at the Roslin we found that four in every ten cloned lambs died within a few weeks of birth – a far higher proportion than would have been lost in commercial flocks in southern Scotland. In fact, the clones did even worse than these discouraging figures suggest, since they had twenty-four-hour veterinary care, unlike their peers on the average Scottish farm. Although we have learned how to take particular care of these unusual animals, the numbers that die just after birth remain significantly greater than normal. In the work by Oback and colleagues, referred to already, only 89 of the 133 calves that had been studied at the time of publication survived to weaning at about three months of age. This proportion – 67 per cent – is far lower than would be expected for animals being monitored as intensively as the clones were. Alarmingly, the groups in New Zealand found that the annual mortality rate in cloned cattle up to four years of age is at least 8 per cent. This is in stark contrast to the long-term survival rate for the offspring of clones, for which there has so far been no mortality after weaning.

Because of the technical shortcomings of many cloning experiments, it has been hard to quantify all the effects, which vary between species. Often it has been impracticable to have an 'ideal' group of animals with which to compare the clones, making it difficult to evaluate the effects of nuclear transfer in a systematic way.

Another problem is that, to boost the chance of success, surrogate mothers are loaded with more cloned foetuses than would result from natural mating, and this may cut their odds of survival if several begin to develop. (Renard's team at INRA reports that a single cloned embryo has about a 7 per cent chance of survival, compared with about 10 per cent when several embryos are transferred. Implanting more embryos does not therefore produce an equivalent increase in the number of clones, because the death of one clone in the uterus can affect the development – and survival – of its brothers and sisters.) Also, scientists often prefer to move on and try something different, rather than dwell on a failure and why it happened. While understandable, this undermines efforts to figure out what is going wrong and why. Finally, there is the possibility that cloning produces abnormalities not seen before and thus these could be overlooked by pathologists who are looking only for what they expect to see.

There appear to be real differences in the precise pattern of abnormality between species and perhaps between different procedures and donor cell types. But the bottom line – cloning does not work very well – is the same regardless of laboratory, cloning technique, species or choice of donor cell. In fact, the overall pattern of problems is so varied that many of us have wondered whether there are any normal clones at all (except that in mice there do seem to be). Even in clones that are apparently healthy, the function of some genes lies outside the range that would be expected if the animals had been produced by normal reproduction. When the epigenetic mechanisms that regulate gene function go awry, the proteins they describe are over- or under-produced. Perhaps a wider range of patterns of gene function is more compatible with normal life than we had previously imagined. However, there may be long-term consequences that have yet to be understood, consequences that may emerge only when they affect something as complex as the human brain.

In the routine meetings between scientists at international conferences, other potential problems of cloning have emerged. We owe the first description of a failure of the immune system in cloned calves to detailed studies by Renard at INRA, who reported the death of a calf cloned from a cell taken from an ear of a calf, itself a clone of an embryo. The calf developed normally for the first six weeks but then suffered severe anaemia: after birth, the number of white blood cells in the animal declined dramatically when it should have increased. They found atrophy of the organs in the lymphoid system, the thymus, spleen and lymph nodes. These findings were possible only because Renard and his colleagues had collected samples from all their calves, and once the animal became sick they were able to go back and make a detailed analysis – diligence not often shown by other groups.

Clones can be left gasping for breath after birth, as happened with one cloned GM lamb born at the Roslin. The lamb was the first in which a precise fragment of the gene that is involved in scrapie in sheep and BSE in cattle, known as PrP, was removed, but there is no reason to think that this genetic manipulation contributed to its fate (BSE-resistant cows have now been cloned in the Far East). At birth it was a fine, strong animal with a healthy appetite and bags of energy. It was removed from its mother to prevent any spread of infection, and was being hand-fed by a member of staff of the Large Animal Unit, to whom it rapidly became attached. However, it panted all the time, even when it did not exert itself.

None of the treatments that were recommended either by vets or by paediatricians could stop it from hyperventilating. We decided that it was kinder to end the lamb's life rather than let it suffer further. Detailed studies by Dolly's pathologist, Susan Rhind, at the University of Edinburgh Veterinary School revealed an abnormality in the blood supply to the lungs of the animal: there was a gross thickening of the muscle that surrounded small blood vessels close to the bronchioles, the small air passages in the lungs. The restricted blood flow at this point would inevitably have made the lamb pant in order to take up enough oxygen and discharge carbon dioxide.

Another common problem with nuclear transfer is unusually big and heavy offspring – 'large-offspring syndrome'. This becomes apparent during pregnancy: gestation takes longer, and there is a sluggish onset of labour and a difficult birth. The larger offspring often fail to breathe after birth; less frequently, there are abnormalities of some organs such as the kidney and the brain. This syndrome had been seen before, when an embryo was grown in less than ideal conditions, and it is now also linked, to a lesser extent, to any method used to handle and transfer embryos to a surrogate mother. It seems that the more procedures that are applied to an embryo, the more damage is caused: even handling embryos causes abnormalities. Growing the embryo in a serum-enriched medium increases them markedly. (Serum was included in early embryo recipes because it was expected to be beneficial. We now know that while serum has some good effects, it actually does more harm than good.)

Even though the embryos of each species are affected to different extents, we already know of a human syndrome, Beckwith–Wiedemann, where size is increased. Because the process of IVF seems (the numbers are too small to be sure) to increase the incidence of this syndrome, it appears that reproductive technologies disturb the finely tuned mechanisms that regulate how genetic instructions are carried out. However, the abnormalities occur most frequently and to the greatest extent in offspring produced by nuclear transfer: cloning wreaks more havoc on these mechanisms than on any others. There have been claims that one day you will be able to spot clones on the beach by their large belly buttons, a relic of being born big with a big umbilical cord. That is, of course, if they survived that long.

Short-Lived Clones

Mice have also provided important clues to that all-important question of how long we can expect clones to live. This comes about because it is possible to clone large numbers of them, to compare and contrast different strains, and draw upon a large number of interrelated studies, for instance on mouse embryonic stem cells. Experiments on mice are uniquely controllable because the mice and cloned mouse embryos that are used are genetically consistent in a way that is impossible to achieve in other species. Furthermore, their housing, environment and diet are very closely regulated. Yet the results for mice are as variable and inconsistent as in any other species.

Only for mice has it been possible to describe in detail the effect of being a clone on lifespan. Because cloning methods are so recent, not enough clones of other species have had the opportunity to live out their natural lives. Mice seem to be more resistant to large-offspring syndrome, but like other animals they have their fair share of other problems. As one would expect, cloned mice do not tend to live as long as their naturally conceived peers, though there are now some known instances of normal physiology, fertility and lifespan.

To explain this result, some have focused on the effects of cloning on a cellular 'timer', called a telomere, which consists of a strand of DNA attached to the end of all chromosomes, rather like the plastic on the ends of a shoelace. Every time a cell divides, the telomeres become shorter. Eventually they run out, and the cells in our bodies stop dividing. This cellular ageing phenomenon is known as the Hayflick limit after its discoverer, Leonard Hayflick of the University of California, San Francisco. We found that with Dolly, although we could turn back time in one sense, converting an adult cell into an embryonic cell, that did not reset the cellular ageing clock, and her telomeres were 40 per cent shorter (older) than those of a typical sheep of her age. But this is not always the case, even in other sheep, and we wondered whether, given that we made the measurements on Dolly's immune cells, she was suffering from an infection that truncated her telomeres.

Later experiments have shown that Dolly was an exception and that telomeres are restored in most cloned animals, so cloning can be 'rejuvenating' in this limited respect. The American company Advanced Cell Technology described the cloning of six calves generated from fibroblast cells at the end of their lifespan, after 1900 attempts. There were the usual problems with cloning – individuals that were unusually large at birth and suffered with breathing problems and high blood pressure. Others appeared normal at two months. The telomeres of these animals all looked like those of newborn calves, ACT reported in 2001 in the journal *Science*, even though they were cloned from senescent cells in which the telomeres have been worn down by ageing.

But it is probably too simplistic to focus on telomeres alone when it comes to ageing. An American team led by Teruhiko Wakayama cloned six successive generations of mice and found that telomere length increased slightly each time. But they lived only as long as other clones, showing that the link between telomeres and ageing is by no means clear-cut. Most likely, premature death was linked with a number of different health problems, such as severe pneumonia, necrosis of the liver and cancer.

In another study, cloned mice became obese in adulthood. If this result extrapolates to people, a cloned human baby will risk growing unusually obese in middle age. Importantly, the offspring of these obese mice were all normal. Thus the problem seems to lie not in their genes but in the way these genes were being used in the body: unlike genetic effects that are inherited by offspring, through the passage of genes, differences in the epigenetic mechanisms that regulate the functioning of the genes are not passed on to the next generation.

Epigenetics: Silence of the Genes

Cloning seems to alter how genes are put to work, which is the focus of epigenetics – the study of changes in the ways that genes are used in the body that happen without changes in the genes themselves. Think of it this way: the genes you inherit are all the possible protein 'notes' that can be played by the instruments (cells) of the body. But each cell has a different repertoire, depending on whether it is an eye cell or a brain cell, say. Not all genes present are functioning in any one cell at any one time.

Research in the past few decades has shown that common, complex diseases such as cancer, diabetes and heart disease stem from collections of changes in genes. This additional layer of complexity is reflected in the wide variation in severity, age of onset, ease of treatment, rate of progression and other factors that affect diseases' variability. Epigenetics plays a role in cancer, for example. Before Dolly the sheep, this mechanism was also thought to explain why the cloning of adult mammals had never been accomplished: the simple tune played by an adult cell could not be wound back to the symphony of developmental possibilities on offer from an embryonic cell. Work by Rudolf Jaenisch, at the Whitehead Institute for Biomedical Research, and others has shown that disruptions in how genes are used in the body are responsible for the deaths and deformities caused by nuclear transfer and other cloning techniques.

There are two ways in which epigenetic mechanisms act: one – called imprinting – is to make sure that, of the two copies of each gene in a developing embryo, only the mother's or the father's gene is used. The patterns of imprints are set during what is called gametogenesis, the development of the eggs or the sperm, which occurs a long time before fertilisation. Imprinting is thought to govern the use of at least seventy-five genes, probably a hundred – and perhaps as many as a thousand. A clone will be healthy only if this fundamental imprint holds true.

Many of these imprinted genes are involved with regulating the size of the embryo – as discussed earlier, in the strange case of Kaguya, the mouse with two mothers. It was thought that nuclear transfer left imprints alone, given that they seem to be set in the ovaries and testes. But we at the Roslin and many other groups guessed, from the existence of large-offspring syndrome, that cloning must disturb imprinting, and thus the process that distinguishes maternal and paternal genes must be vulnerable to disruption in some way after fertilisation. Later studies of cloned sheep by Lorraine Young (now at Nottingham University) and her collaborators demonstrated that one gene affected by nuclear transfer comes from the mother – a growth inhibitor called IGF2R, referring to how the gene makes a protein receptor which clears a paternal growth signal, called IGF2, from the circulation. As a result, the less IGF2R, the more IGF2 and the larger the offspring grows.

Recently, scientists from the United States, Sweden and Japan implicated the same mechanism in tumour growth, where an 'epigenetic' malfunction turns on the copy of the IGF2 gene that should remain off. Having a double dose of the IGF2 protein, as do one in every ten people, is sufficient to change the normal balance of cells within the lining of the colon, thereby doubling the risk that a cancer-causing genetic mutation will trigger a tumour there.

A second epigenetic mechanism, which uses the same molecular mechanisms as imprinting, gives each cell in the body its identity. This ensures that, although every cell in the body knows the entire genetic repertoire, only relevant genetic tunes are played: those required to make a skin cell, brain cell, or whatever. This 'molecular memory', which enables a cell to remember what part of the genetic code to use to maintain its identity, is common to all creatures. That is why a brain cell stays a brain cell and does not turn into a liver cell, even though their complement of genes is the same.

Because the mammary cell from Dolly's 'mother' used only the relevant parts of the genetic code to make a mammary cell, it had been thought that this would block the development of all the cell types required for a clone. Obviously, this memory was erased by the cloning technique used to breed Dolly. But, of course, cloning is risky and inefficient. The reason is that the pattern of gene activity controlled by epigenetic mechanisms is not reproducibly wiped clean by cloning: you may not erase the entire 'molecular memory' of its being, say, a skin cell to obtain an embryonic cell with the potential to develop into any of two hundred or more cell types. If this is faulty, some genes will be used inappropriately and others not at all.

A chemical mechanism called methylation is one kind of gene switch used in epigenetics: decorate a gene with methyl groups, and you turn it off – the cell no longer plays that 'note'. Normal development of an embryo therefore depends on the methylation state of the DNA contributed by the sperm and egg, the parental imprints, and the changes in methylation of DNA that help to tell an embryonic stem cell to turn into a skin cell, or whatever. And the workings of cells are, to be sure, vastly more complicated than this. Some scientists even suggest that the genetic code of DNA is joined by a second code, called the histone code, named after proteins called histones, which govern how the body interprets genes. It could well be that understanding the nuances of this mysterious code, in addition to imprinting, can shed important light on what goes wrong in cloning.

The histone code is based on how the 2.3 metres or so of DNA in each cell is bundled up within chromosomes and, more importantly, how it unpacks. Here, the popular picture of chromosomes is muddled and misleading. The X shape used to depict chromosomes in many accounts actually shows the chromosomes during the process of cell division, when they are copying themselves. For the rest of the time the chromosome would look more like a sausage with a pinch in the middle. But perhaps it is better to forget chromosomes altogether, given that we are most interested in what happens when the cell is reading the genetic notes on the coils of a single DNA molecule bundled into a chromosome. If you can see the chromosomes, then the cell is going through cell division, and the DNA wrapped up inside them is doing little more than contributing to this process. Only when a chromosome unravels, to leave a spaghetti-like mess of DNA in the nucleus of a cell, does the serious business of playing a genetic tune begin.

The ravelling and unravelling of DNA is controlled by proteins. In fact, chromosomes are made of a blend of the double helix of DNA and protein that together are called chromatin. In one form, dubbed heterochromatin, the genes are all packed away and out of use. This is the stuff that the visible regions of the chromosomes are made of. In the second form, called euchromatin, the DNA spills out into the interior of the nucleus so that the genes can be put to use – played, to continue the musical metaphor – to make proteins that can build and operate the cell.

A conductor's-eye view of a cell can reveal that, in the apparent chaos of this DNA mess, there is order. One might expect that when the chromosomes unpack so that their genes can be used, the nucleus of a cell would be filled with a tangled mass of DNA from all the chromosomes (forty-six in humans). Instead, one discovers that the chromosomes are always found in the same regions in the nucleus, even after the process of spaghettification. Jane Skok of University College London studies the position of the genes encoding antibodies in certain white blood cells (called B cells) by means of DNA probes that are able to bind to them. By analysing the position of these genes in individual cells, she is able to see that in a cell where a gene is being used, the DNA that describes it is somewhere in the middle of the nucleus. When the gene is out of action, it is tucked away on the periphery of the nucleus.

Of the various proteins that control how the message of DNA is read, perhaps the most important are the histones. Certainly they are the best understood. Think of a cheap string of pearls, where each pearl is a spool around which DNA is wound and the elastic is the DNA. Around a ball of eight histones the DNA is wrapped two and a half times to form a 'pearl', called the nucleosome. The cell reads the DNA's message rather in the manner of groping for the pieces of elastic between the histone pearls. In a chromosome, the nucleosomes are packed tightly together, like a necklace that has been wound up so that it can be easily put into a purse.

To control the packing and unpacking, and thus the way that genes are used, the histones in each nucleosome have a 'tail', dangling into the nuclear soup like a drawstring for the necklace. Chemical modification of this tail controls how genes are used: changes can separate the pearls to expose DNA strands between them, move them around the DNA, or pack them together and so lock genes away. If a process called histone acetylation takes place, genes tend to go into action: the nucleosomes move apart to allow the DNA to unravel so that certain proteins in the soup of molecules swimming around in the nucleus can go to work on the genes that link the loosened nucleosomes and carry out their instructions. Histone methylation (not to be confused with DNA methylation) acts in two ways: depending on where the methylation mark is positioned, it can turn genes either off or on.

The histone code is crucial for development. Compare a body cell (a somatic cell) with a germ cell, such as a sperm, and you will find that the chromatin structure is quite different. In the few hours after a reprogrammed embryo is activated, important changes to chromatin take place. At least one histone has a form that appears only in the embryo. All this suggests, as one would expect, is that turning a spherical fertilised egg into a foetus requires a lot of changes to gene regulation. Studies of cloned mice show that changes to chromatin as well as changes to DNA methylation can be messed up by nuclear transfer. And as a result of this 'nuclear remodelling', some genes are turned on when they should be turned off, and vice versa. The embryo dies. The foetus fails to develop normally. An offspring is born with serious abnormalities.

To prevent these problems from occurring, we need to know more about how genetic material is used in the first, critical twenty-four hours after an egg is fertilised. This is now much more the focus of my interest than nuclear transfer itself. It should already be apparent that the regulation of genes is highly dynamic and complex: it is going to take a long time to understand these epigenetic mechanisms, to know what normally happens, and then intervene during cloning to fix what nuclear transfer fails to achieve. At present we expect too much of eggs in reprogramming adult DNA and have to give nature a helping hand. We need to improve nuclear transfer.

Beyond Dolly

Efforts are under way to hone nuclear transfer in various ways. The Dolly method involves highly precise manipulations to remove and inject DNA under a microscope, using apparatus costing tens of thousands of pounds. Of the many variations on this theme that have since emerged, I have come to prefer two quicker and easier alternatives, called zona-free methods, which are less reliant on expensive equipment. One was developed by Teija Peura and colleagues from the South Australian Research and Development Institute in Melbourne, and refined by Gabor Vajta and his colleagues in Denmark. The second was developed by Björn Oback and colleagues at New Zealand-based AgResearch, and the University of Waikato.

In the first zona-free method, Peura uses a short cut in enucleating the egg. Rather than hunt for the chromosomes and remove them specifically, the egg is simply cut in half with a hand-held blade (this was the approach used by Steen Willadsen, as discussed in Chapter 3). Ultraviolet light is then used to identify and discard the half that contains the chromosomes. Two halves of eggs emptied this way are put back together again, like Humpty Dumpty, and nuclear transfer is achieved by fusion of the donor cell with this recon-structed egg. Although it takes two eggs to make each clone, the reconstructed embryo is likely to be slightly larger in volume than those that resulted from the Dolly procedure, and thus to have more cellular machinery in the cytoplasm upon which to draw and there-fore a better chance of developing well. One potential problem with this method is that the resulting embryo contains mitochondrial DNA from two different eggs.

The second zona-free method, developed by Oback, uses micro-manipulators in a way that is much quicker to learn and carry out. However, it does not offer any increase in efficiency. Enzymes are used to strip the egg of its zona pellucida. The naked egg is then held against a pipette, rather than being sucked against it, as in our procedure. Ultraviolet light is used to guide enucleation with another pipette, a process so quick that there is little damage. Then a donor cell is positioned next to the naked egg in the presence of lectins, sticky compounds that glue the two together, and the two are fused with an electrical pulse. In the Dolly method, the egg and the tiny donor cell had to be aligned exactly by hand between the electrodes for fusion to occur. This alignment is essential if the two cells are to fuse efficiently. Strikingly, in this zona-free method the electric field does all the hard work and lines the two up properly – the egg and donor cell roll around and align themselves so long as the zona is not there. This technique takes an inexperienced person just a few weeks to learn, compared with six months for the origi-nal Dolly method. Even though it does not increase cloning efficiency, it effectively doubles the throughput in cloned embryo and cloned offspring production. Because of such advances, we can see the day when robots and computers will carry out the delicate manipulations required for cloning.

Flash of Inspiration

Something else that could make cloning safer is an easier way to achieve activation, the moment when a flash of calcium triggers development. When this happens naturally, one can detect pulses of high calcium concentrations every few minutes that continue for several hours. There are various methods of triggering a surge of calcium in cloning, for instance by applying an electrical current or even by giving the reconstructed eggs a calcium bath. But most methods generate one very large wave of high concentration calcium for hours. The development of successful procedures for nuclear transfer in a species for which it has not been tried before has involved the adaptation of the method of activation to suit that species. To be successful, it is crucial to get the details right.

To improve cloning in general it would be good to try to mimic the natural waxing and waning of calcium after fertilisation. A group in France led by Jean-Pierre Ozil has spent years working on a box of electronics to create precisely the same sequence of calcium pulses that occur during fertilisation, giving eggs bursts of calcium in a tiny bath. They have shown that it is possible to increase the success rate of producing parthenotes, but their method has not been incorporated into routine cloning procedures because it is still a lab prototype and not yet commercially available.

An alternative strategy is to find out more about how activation occurs naturally and to identify and use the particular protein in the sperm that triggers the development of a fertilised egg. The search for the sperm signal that awakens the egg began two centuries ago, and knowing what it is is vital for understanding how a fertilised egg starts life anew. Recently, this trigger was revealed to be a protein – PLC–zeta – by Karl Swann of University College London and Tony Lai at the University of Wales College of Medicine. This protein switch could offer a sophisticated alternative to currently used activation methods.

Cytoplasmic Engineering

An alternative way to get development – from imprinting to the histone code – off to the right start in nuclear transfer is to pretreat the donor nuclear DNA in some way. One approach, developed by John Gurdon, is to have faith in how nature uses very similar machinery to do important things in a cell, whatever the species. At a particular stage in frog egg development (frog eggs offer plentiful cytoplasm compared with tiny mammalian eggs), he injected in a mammalian cell that had previously had holes made in it, allowing it to set up a 'chemical dialogue' with the frog cytoplasm. Through this experiment, done in the 1970s, he could tell that some of the mammalian genes were turned off and some egg genes were used. More recently, he has shown that one gene (Oct4), known to signify embryonic stem cells, goes into action. In collaboration with colleagues at the Roslin and Julian Blow of the University of Dundee, we are investigating how a similar approach could help to reprogram the DNA used in nuclear transfer.

Complementing this is the use of technological advances in genomics to screen the patterns of tens of thousands of genes to identify developmental differences between reconstructed embryos and those produced naturally or by IVF. This study will be crucial for the efforts to see how assisted reproduction techniques can make genetic melodies played by each cell drift off key or off tempo to disturb development.

There is another interesting feature of the first twenty-four hours of life. The parental DNA has not been mixed to form an individual at this stage, and the DNA is not even being used (the details of how RNA and DNA are used in the zygote vary from species to species), as it is in a normal cell. We need to know how parental DNA is organised at this critical time. And we have to know about the parental genetic messages scurrying hither and thither to control and influence the early stages of development. These messages are written in RNA, the more ancient genetic code found in our cells alongside DNA, and are conveyed via proteins to the zygote (early embryo/placenta), mainly by the egg, but a small number are contributed by the sperm. One reason why eggs work at the MII stage of development is that they possess the right RNA instructions to help direct protein manufacture and development, and also perhaps to organise the chromatin. Although their exact function is unknown, it is the RNA codes that guide the development of the early embryo, before it starts to use the instructions in its own genetic code – just as you need an operating system to run a program on your computer. One can think of this as an RNA operating system that enables the development software to run in the cell.

One experiment that shed light on this hitherto unknown aspect of our inheritance was performed by Stephen Krawetz of Wayne State University School of Medicine in Detroit, and colleagues in Britain, which revealed that men play a more important role in procreation than they may have thought. Despite cloning and Kaguya, the first fatherless mouse, Krawetz found that sperm do more than simply deliver a father's DNA to an egg. In human sperm and in fertilised eggs there are six messenger RNAs, each corresponding to a paternal gene, that are not present in unfertilised eggs. This suggests that sperm deliver at least six RNAs – probably more – to the egg at fertilisation. Krawetz has also found micro-RNAs, a recently discovered large class of regulatory, non-coding genes that bind to sites in target messenger RNA to regulate how they are broken down and used in the zygote. In humans, sperm also deliver a structure called a centriole, a key part of the machinery of cell division. These findings could have implications for the success of cloning, where only DNA is transferred. Perhaps, by adding a sprinkle of paternal RNA as well, the low success rate could be boosted for research to clone embryos as a source of stem cells for treatments.

Another focus of efforts to improve on the Roslin technique rests on a feature of Dolly that is not widely appreciated: although the process of cloning exchanges the nuclear DNA of an egg for that from a donor, it leaves mitochondrial DNA from the egg. Thus Dolly had the nuclear DNA from a mammary cell from a Finn Dorset sheep and the mitochondria of the egg, which came from a Scottish Blackface. Strictly speaking, Dolly was not a clone because of this difference in mitochondrial DNA, but most of us will continue to use the familiar term when referring to her.

Dolly therefore relied on genetic material from the Blackface – around twenty genes in the donor egg's mitochondria. While the amount of information in the mitochondria is small in comparison with that in the chromosomes, it is still essential for normal development. Abnormalities in mitochondria, which can accumulate with ageing, can lead to disease. That goes for human clones too, where in most cases the egg that was used during nuclear transfer will have been obtained from a different woman from the one who donated the DNA – if the latter is a woman at all. The relatively low expectations we have of a fertilised egg, whether it is in the laboratory or in the womb, reflect the important differences between eggs in their ability to support the normal development of an embryo. These variations in the success of development can be down to differences in the mitochondria, and may mean the difference between the life and the death of an embryo. We need to understand more about what causes them and how to tinker with these factors if we are to improve success rates.

The Folly of Reproductive Cloning

There are some people, of course, who want us to go further and take human cloning to term. They have an easy answer for the difficult problems I outlined above: simply select the good embryos, the ones for which there is no evidence that gene regulation has gone awry. This suggestion rests in a misleading way upon the fact that it is already possible to identify those embryos that have inherited known genetic abnormalities from a parent, thanks to the technique of pre-implantation genetic diagnosis. With cloning, however, we are looking not at faulty genes but at faults in the way genes are used, caused by errors in the histone code, methylation, or whatever. And to make matters worse, we don't understand the details – in fact, our understanding of the subtleties of gene regulation is pitiful. One cannot devise a test for a poorly understood problem.

To try to clone a baby using current methods would be a terrible gamble. Imagine tossing five coins and getting five heads or five tails. That is an unlikely outcome, and yet for a clone to survive and be normal would be about as likely as seeing five heads. It seems self-evident that it would be grossly irresponsible even to consider using the present procedures to produce a cloned child. The likely outcomes of any such attempt would include late abortions, the birth of dead children and, perhaps worst of all, the birth of children who survive but are destined for a lifetime of pain and disability. Remember the cloned lamb that panted all the time because of an abnormality in the lungs? It is distressing enough when an animal suffers this way, but it would be awful to have inflicted this fate on a human child.

One is tempted to conclude that reproductive cloning should never be tried until proved totally safe. But, as we all know, absolute safety is impossible, and if we used this simplistic rule we would still be living in the dark ages: with the benefits of every new technology always come risks. Yet it is true to say that, for now at least, reproductive cloning must be banned because it is too risky. The problems look daunting. But I would never go so far as to say that they will never be overcome. Everything rests on details of the technique, such as the use of fresh eggs. I have no doubt that this kind of study, and those that will attempt to demonstrate that cloned embryonic stem cells are normal, will help to show us how to deal with many of the problems facing cloning to produce children. Moreover, I believe that there will be no obvious application, because there will be alternative ways to meet apparent needs. Specifically, I believe that it will be possible to produce eggs or sperm for infertile individuals. Several groups have now shown how embryonic stem cells can develop in the laboratory into the early forms of cells that eventually become eggs or sperm. If the stem cells were derived from a cloned embryo, then the eggs or sperm could be used by the person from whom the nuclear donor cell was taken.

In early work with mouse embryonic stem cells, Hans Scholer's team at the University of Pennsylvania used embryonic stem cells to make cells with some of the characteristics of eggs. Toshiaki Noce of the Mitsubishi Kagaku Institute of Life Sciences in Tokyo has already pushed mouse embryonic stem cells towards sperm production. George Daly and his colleagues at the Children's Hospital Boston/Harvard Medical School and the Whitehead Institute for Biomedical Research have injected immature mouse sperm created in this way into eggs to make embryos. Their work also opens up the possibility that eggs and sperm could one day be grown from stem cells and used to study the effects of hormone-disrupting chemicals, for assisted reproduction, therapeutic cloning, and the creation of more stem cells for further research and for the improved treatments for patients suffering from a range of diseases.

Although it remains to be seen which of these more speculative strategies will eventually pay off, these techniques offer many new opportunities for infertile couples to reconstruct their eggs and sperm to have children that are genetically their own, who will contain a blend of their own genes, as would result from making a baby in the usual way. Cloning, meanwhile, offers only the unusual option of having an identical, and younger, twin of one parent. In short, I feel confident that there will be many alternatives to the use of reproductive cloning, and that it will not be necessary. Nor do I feel that society will ever be comfortable with reproductive cloning. Attempts to clone offspring will be condemned as unethical.

Beyond Human Cloning

Like many of my colleagues, I believe that the prospect of routinely growing cloned embryos so that cells and tissues can be grown for patients is unlikely on a large scale or in the long term, though crucial in the short term. It is crucial because we know that this class of cells, above all others (notably adult stem cells), has the wherewithal to develop into any type of cell. If we can perfect this process and make it reliable, then we will not only obtain cells for repair and for testing drugs, but also gain profound insights into the details of development and disease. And on the basis of information from this research on therapeutic cloning, and insights from the study of more differentiated 'adult' stem cells, we can devise embryo-free methods that can wind back the developmental clock of a patient's cell so that it can recover its full genetic potential. If we can understand what happens in the wake of nuclear transfer, we will one day be able to mimic those events and thereby actually change cells from one type to another – so that after a heart attack, say, skin cells from the patient might be turned by a cocktail of factors into heart muscle cells and used to repair the damage to the heart.

Immunologically matched cells would thus be derived without the expense and difficulty of cloning. We would be able to grow cells of any type to repair a patient without the need to create human embryos – and, of course, without inviting the disgust of those who still believe that life begins at conception.

There are already hints that this feat of cellular alchemy is possible. In 2002, Philippe Collas of the University of Oslo's Institute of Medical Biochemistry reported how he gave one type of body cell – fibroblasts – some of the characteristics of another type, white blood cells called T cells. His team made holes in the cell membranes of the fibroblasts and dunked them into an extract from the T cells. This reprogrammed the fibroblasts to lose some fibroblast functions, but to gain some functions from the T cells. Since then, Collas has converted fibroblasts into insulin-producing cells, albeit transiently, and fat cells into the cardiomyocyte cells found in heart muscle.

Another approach has been taken by Peter Schultz and Sheng Ding at the Scripps Research Institute in La Jolla, California. They have been screening molecules in a hunt for ones which can turn the clock forwards or backwards in a cell's development. The chemists are on the trail of a small molecule they call 'reversine' which can cause a muscle cell to turn into a progenitor cell that has more developmental potential, and could be turned into a wider repertoire of tissue types.

Ultimately, scientists want to use these methods to persuade body cells to 'dedifferentiate' – to make the journey back to earlier, more plastic stages. Thus cells of any type could be regenerated from a speck of a patient's tissue for a vast range of treatments and repair. This approach would satisfy the main promise of therapeutic cloning: the creation of cell lines tailored to a patient which avoid the problems of immune incompatibility. It would also eliminate the need to obtain donated human eggs and, most importantly, it would avoid attracting the moral outrage of those who oppose the use of human embryos.

However, one other application of nuclear transfer still involves embryos. I think that it falls between the extremes of reproductive cloning and therapeutic cloning. This clone is not of a person, but of a blastocyst. And, unlike reproductive cloning, this combination of genetic modification and cloning seems to me to offer an important way to reduce human misery, and therefore deserves consideration.

Chapter 9

Designer Babies

Polly the sheep has something important to say about the future of humanity. The combination of nuclear transfer and genetic modification used to create her has broad applications in human medicine that could prevent a great deal of suffering. They may help to prevent the birth of children with devastating genetic diseases. Because public attitudes to *in vitro* fertilisation have shifted from alarm to acceptance, I can even foresee the day when the genetic modification of human embryos is accepted as routine, even if it is not used on a large scale. When it is possible to produce genetically modified people safely and predictably (which is still a distant prospect), humankind will take on more responsibility for its own evolution.

But I can see profound ethical issues arising from what some condemn as 'playing God'. Patients, parents and regulators will bear a heavy burden of responsibility when it comes to deciding whether this technology should be used and, if so, in what circumstances. Some believe that the desire to prevent the birth of disabled children is often driven by misinformation and prejudice. Others see this as another unwarranted intrusion into natural procreation. Many fear the subtle long-term effects on society: would a child's sense of independence or self-worth change on learning that his parents had intervened in his birth to change them in some way? These technologies could also be used to make someone not just well but better than well – to expand muscles, enhance memory or delay ageing.

I am not working on this technique – at least not at the moment – but I believe that it is important to discuss it because of its potential impact, from benefits to pitfalls, and the difficult choices that lie ahead for individuals as well as for society. However, there is much confusion about exactly what scientists propose to do, what they are actually capable of doing and whether they should be allowed to do it at all. Several techniques are already on offer to deal with hereditary disease, such as gene therapy.

Gene Therapy

The most advanced technology that we possess to tackle the root cause of genetic disease is gene therapy – gene transplants, if you like. This therapy can be used, for example, to correct the damaged genes responsible for diseases such as sickle-cell anaemia. Sickle-cell disorders are most common in individuals of African, Mediterranean, Indian and Middle Eastern descent, and one in every thirteen African–Americans carries the sickle-cell trait. The disease is caused by a single genetic 'spelling mistake' – a single nucleotide mutation – in the human beta globin gene that leads to an abnormal form of haemoglobin, a protein that can capture and release oxygen.

One copy of the sickle-cell gene is beneficial, providing an increasing amount of protection against malaria during the first decade of a child's life, according to a study by Tom Williams of the Kenya Medical Research Institute/Wellcome Trust Research Programme in Kilifi. However, children who inherit a double dose of the sickle gene suffer from sickle-cell disease, which in its worst form causes chronic ill health and can result in early death.

Red blood cells supply oxygen to the body by means of their cargo of haemoglobin, which normally floats freely within them. But sickle-cell patients have a mutated, 'sticky' form of haemoglobin that tends to clump together into long fibres. The fibres form a scaffolding that distorts the cells into a sickle shape, so that they then clog small blood vessels. These traffic jams deprive vital organs of oxygen, and patients end up with anaemia, jaundice, major organ damage and many other maladies. Theoretically, gene therapy could cure this disease.

To do this, scientists would take a corrective gene – a normal form of the beta globin gene – and introduce it into the bone marrow cells of patients. The job of genetic surgery can be done by a modified virus, a complex chemical that has been honed by nature to insert its genetic code into cells and pirate their cellular machinery. Rather than convert a cell into a virus factory, the virus is modified so that it inserts a therapeutic gene. Another approach is to package the gene in a fat particle, called a liposome, or even to inject naked DNA.

Gene therapy has been tested many times. The first approved human trial of gene therapy was carried out by Michael Blaese and colleagues in September 1990 at the National Institutes of Health in Bethesda, Maryland. At the time the media hailed this experiment as the dawn of a medical revolution that would see treatments for the four thousand or so hereditary diseases that are caused by a defect in a single gene, and even for cancer, heart disease and other ailments where genes play an important role. But because the two young girls – sufferers of a rare immune deficiency – who received the gene therapy also continued to receive conventional treatment, the results of that pioneering trial were ambiguous.

Although studies on animals continued to suggest that gene therapy had great promise, human trials during that decade were disappointing. A glimpse of the true potential emerged only in 2000, when gene therapy produced remarkable results. A paper in the journal *Science* described how the therapy had released three boys from imprisonment in a germ-free jail. Alain Fischer and colleagues at the Hôpital Necker in Paris treated the baby boys, aged from six to twenty-one months, who were affected by a genetic disorder that causes a severe immune deficiency which strips away the body's defences. Called severe combined immuno-deficiency (SCID) X1, the X referring to how it is inherited via the X chromosome, the disease forced the boys to live within sterile 'bubbles' (children born with acute SCID are known as 'bubble babies') to protect them against the serious threats from, for instance, just a cold sore or chickenpox. They would die within a year without a bone marrow transplant to rebuild their non-existent immune system, but this transplant works in only around 60 per cent of SCID-X1 patients and necessitates additional treatments.

Instead, thanks to gene therapy, the boys were given a normal copy of the defective gene responsible for the disease. First their bone marrow was removed. Then the stem cells in the marrow that are the 'parents' of the cells that form the immune system were enriched, infected with a mouse virus containing the corrective gene, and finally returned to the body. The lives of these boys were changed out of all recognition by this genetically altered mouse virus (referred to as the vector). Previously, the boys had had severe diarrhoea and been unable to eat because of chronic gut infection, so they had to be fed intravenously. After gene therapy they had a healthy gut and could eat a normal diet for their age. No longer in protective isolation, they were able to live at home without any treatment, enjoying normal growth and development. This transformation was remarkable, and success was also reported in similar trials by Adrian Thrasher and Bobby Gaspar at London's Great Ormond Street Hospital.

But the brilliant record of the French team would become tarnished. Almost three years after the treatment, white blood cells in the two youngest patients began proliferating abnormally, as happens in the blood cancer leukaemia. Subsequent investigation showed that the French vector had turned on an oncogene – a cancer-causing gene – called LMO2, which is required for the production of blood cells. One child died and, at the time of writing, a third had developed to leukaemia. Despite this tragic setback, patients in the British trial have not suffered serious side effects. Gene therapy is still thought to offer a safer alternative to bone marrow transplants from a parent or mismatched, unrelated donor. And the vectors – viruses – are being made safer and more effective all the time.

Apart from the dangers, which can arise with any experimental technique, another consequence of this kind of gene therapy should be highlighted. The direct influence of the transplanted gene is confined to the boys alone. The genetic repair carried out by the French and British teams would not be inherited by any children that the boys might subsequently father, because the gene transplant affected only the bone marrow. This is what is called somatic gene therapy – where somatic cells are non-reproductive cells – and this limitation is crucial for ethical approval, since there was, and remains, unease about using any germ line gene therapy which would pass the new gene on to a recipient's descendants. The human gene pool would be permanently affected as a result, and some even talk of 'genetic pollution'. Although these changes would presumably be for the better, the fear is that an error in technology or judgement could have far-reaching consequences.

The Story of ANDi

Gene therapy can also be carried out at a much earlier stage of life to modify all the cells in the body, including germ cells. This was attempted in the creation of the world's first genetically modified primate, a baby rhesus monkey called ANDi ('inserted DNA' spelt backwards). Gerald Schatten, then at the Oregon Regional Primate Research Center in Portland, led the group that added an extra gene, one that codes for a protein called jellyfish–green fluorescent protein. This protein was of no medical interest but was very easy to detect – it makes the recipient glow green under ultraviolet light. The success of ANDi would give the green light, as it were, for the creation of primate models of diseases such as Alzheimer's.

Leaving aside the vexed issue of whether primates should be 'humanised' for animal research, the risks taken in creating ANDi would be unacceptable for use on people. Schatten and Anthony Chan used a virus containing the gene in the form of RNA code to genetically modify 224 eggs, and then fertilised them to produce forty embryos. The embryos were transferred to twenty surrogates, but only five pregnancies resulted. Three healthy male babies were born, two infants were stillborn and another pregnancy did not develop. Of the healthy infants, only ANDi contained the new marker gene – revealed when the team amplified and analysed genetic material extracted from the inside of his cheek, his hair and cells in his urine, as well as studies of his placenta and umbilical cord. Even so, although ANDi took up the gene, the experiment was not really a glowing success: the gene did not seem to be in use in ANDi's cells.

The experience of the French team with bubble babies and of the Americans with ANDi seems very familiar to someone like me who has spent years altering the genetic make-up of mice, sheep and cattle. We have all found that it is possible to make genetic changes, but there are serious limitations. First, it is possible only to add genes, not subtract or repair them. Second, the efficiency is so low that only a small proportion of the offspring carry the additional gene. Finally, the genes are not always used in the body. And, as the French team showed, even if new genes are successfully taken

up, they can cause problems. Quite clearly this is a technique that
has to be greatly refined before it can be put to routine use in
human medicine. However, an alternative way to deal with hereditary disease that does not require genetic tinkering is already in
widespread use.

Selecting Embryos

In the old days, couples who were aware that they had a history of
a genetic disease in their families but still wanted children had no
option but to go ahead not knowing whether any or all of their
offspring would inherit the disease. For them, life was a lottery
with very high stakes. Take the example of haemophilia A, the most
prevalent and serious type of haemophilia. Unlike Christmas disease, which we encountered in Chapter 5, this more common
form of the disorder is caused by a mutation in the gene responsible for factor VIII, a protein made mainly in the liver that plays a
role in clotting. The mutation can arise spontaneously, as it did with
Queen Victoria, who reigned over Great Britain between 1837 and
1901.

This tiny DNA alteration sent ripples through history. Victoria's
eighth child, Leopold, was affected and died aged thirty-one from a
haemorrhage. Two of her daughters, Alice and Beatrice, were carri-
ers and passed the faulty gene to several royal families in Europe,
including those of Spain and Russia. In the Romanovs, the gene
passed on to Victoria's granddaughter, Princess Alexandria. The
princess was married to Czar Nicholas of Russia, and their son, the
czarevitch, was also affected with the disease. The princess and the
monk Rasputin, who treated the czarevitch, had considerable political influence, and their machinations may have been responsible in
part for precipitating the Russian Revolution.

One step towards ending this kind of deadly lottery came with the development of various blood tests that could help to check the health of the unborn child. For example, there is a blood test that can be used by doctors to estimate the risk of the foetus having certain defects, including spina bifida, Down's syndrome, and other chromosomal abnormalities or structural defects. In the case of Down's syndrome, caused when the foetus has an extra copy of chromosome 21 and thus an overdose of the 225 genes that it carries, levels of two particular proteins in the blood of pregnant women have been found to be an indicator of risk. The unborn baby is considered to have a high risk of contracting the syndrome if tests reveal that the free beta human chorionic gonadotropin (a hormone made in pregnancy) is present in high concentrations, and pregnancy-associated plasma protein A in low concentrations. Foetuses with Down's syndrome also tend to have an excess of skin at the back of the neck (nuchal translucency), which can now be detected by an ultrasound scan.

To make certain of their diagnosis, doctors use amniocentesis and chorionic villus sampling (CVS), both of which rely on obtaining foetal cells so that their chromosomes can be analysed for defects. Amniocentesis has traditionally been performed at fifteen to eighteen weeks' gestation, when a needle is used to collect fluid from the amniotic sac. In CVS, which can be done as early as ten to twelve weeks' gestation, a sample is taken from the mat of chorionic villi, the finger-like projections on the foetal side of the chorion frondosum, the early placenta. Both tests carry a risk of miscarriage, so usually they are used when other tests suggest that there may be a problem.

Alternative tests, which exploit the discovery that genetic material from the baby, such as messenger RNA, circulates in the bloodstream of pregnant women, offer the prospect of a safer and earlier way to screen the foetus for genetic disorders. Whatever the test, however, if it reveals that damaged genes are present, then prospective parents are faced with the choice of either continuing with the pregnancy to produce a child known to have inherited the genetic disease, or terminating the pregnancy. A dreadful dilemma.

New hope for families with a history of genetic disease came in 1990, when Robert Winston, Alan Handyside and colleagues from Hammersmith Hospital in West London announced that they had developed a way to test IVF embryos so that only those that were free of a serious defect were implanted in the mother. They carried out a genetic test on DNA amplified from a single cell taken by a needle from the early embryo, at the time when it consists of a ball of six or eight cells and is only a tenth of a millimetre across. Fears that this biopsy is likely to increase the risk that the embryo will not go to term do not seem to have materialised: pregnancy rates after this method of pre-implantation genetic diagnosis, or PGD, are the same as for normal IVF.

The team initially used this method to screen the sex of embryos from parents who carry disorders that would affect only boys, such as Duchenne muscular dystrophy, a fatal cause of muscle wasting, and childhood adrenoleukodystrophy, a devastating progressive disease best known from the case of Lorenzo Odone, whose story is told in the 1993 film *Lorenzo's Oil*. That way they could ensure that only female – and thus healthy – test-tube babies were transferred to the mother. 'For the first time,' said Winston, 'people who are carrying a severe genetic disease can start a pregnancy from the beginning, knowing it is normal, and do not have the spectre of a termination of pregnancy.'

Since then, PGD has been extended to the analysis of embryo cells and the polar bodies cast off either before or after fertilisation, and broadened to help tackle many other hereditary diseases, to screen out genes that raise cancer risk, and to rhesus factor disease – the potentially fatal condition caused by incompatibility between a baby's blood and that of its mother. The technique will soon be able to screen for any defect as easily as can conventional methods, such as amniocentesis, thanks to new methods that amplify all the DNA in the cells taken in the biopsy, rather than just the DNA linked with one specific disease, a method called whole genome amplification. The method can also be used to check the number of chromosomes, since chromosomal abnormalities play a major role in miscarriage.

The latter method, 'full chromosomal analysis', was developed by Dagan Wells and Joy Delhanty at University College London Medical School. The technique combines whole genome amplification and comparative genomic hybridisation, which detects chromosomal imbalances, to count the number of copies present of every single chromosome in the majority of cells of a pre-implantation embryo. Strikingly, the technique revealed that most human embryos contain a proportion of abnormal cells, some of which arise in the egg or sperm and others from early development, so that by the age of three days 60 per cent of embryos contain a mosaic of normal and abnormal cells. For those who hanker after a 'normal' or 'natural' birth, this should give some pause for thought. In terms of numbers, a 'normal' embryo is chromosomally abnormal.

As ever, PGD raises a number of issues. First, there are questions about the safety of the IVF and embryo manipulation required for PGD. Although hundreds of babies have been born after embryo biopsy with no obvious harmful effects, we need to see the results of a more systematic follow-up of their health over decades to be sure. There are also uncertainties about the accuracy of the genetic tests that are used. Some find PGD, or its uses, morally unacceptable because it involves the creation and selection of human embryos. There are issues about whether it should be used to select against susceptibility as well as disease, and whether it is right to choose a child's sex or other desirable characteristics (though attributes such as musical ability and intelligence are so genetically complex they would seem beyond PGD's capabilities). Other issues have to do with whether a couple who are blind or deaf, for example, can use PGD to select an embryo with a matching disability. Then there are questions of equity: many cannot afford PGD, and it is not clear whether and to what extent health insurers and national health systems will cover the costs. And there are also broad ethical questions about the impact of PGD on family relationships, people living with disabilities and society as a whole. For what it is worth, I feel that, under the kinds of control we have seen in Britain, and with proper advice given to parents who want to undergo PGD, the technology is very unlikely to cause harm and offers many benefits. It certainly provides a kinder alternative to getting pregnant, having amniocentesis and then, if tested positive for a serious disease, having an abortion.

Most controversially, the technique has been used to create what pro-life campaigners, to signal their disdain, call a 'designer baby'. This pejorative term was presumably meant to convey the (misleading) impression that parents can now choose embryos as casually as they select an item of designer clothing. But to me, embryo selection does not count as embryo design. Nor does it seem very different in principle from the selection of healthy looking embryos for IVF. In one highly publicised case, campaigners challenged the authority of the HFEA to license the use of the method to create a 'saviour sibling', a healthy embryo with the correct genetic match to allow cells from its umbilical cord to be used to treat an older brother. In this case the older sibling had the serious blood disorder thalassaemia, and the aim of the exercise was to restore his supply of normal red blood cells. Approval was eventually given, after a court battle. However, even with these formidable legal obstacles overcome, no one could guarantee success. In the first attempt, fifteen embryos were produced. Only one proved to have an exact tissue match, but it carried the beta thalassaemia disease. In the second, ten embryos were produced. Two of these proved disease-free and to have a tissue match. One was implanted, but no pregnancy resulted.

Creating a 'designer baby' is difficult. There is a limit to the number of eggs that a woman can produce at one time. Not all the eggs will undergo fertilisation and proper development. Even if a healthy embryo is successfully tested, PGD does not work every time. Only a fraction of those embryos that are successfully tested will have the 'right' genetic make-up. And IVF is a hit-and-miss affair, and more often than not an attempt to implant the embryo will be unsuccessful. Success rests in a fraction of a fraction of a fraction of the fraction of the eggs that are fertilised. There is no way to guarantee that even one IVF embryo can both be free of the damaged gene and have the right tissue type and successfully develop in the womb. There is a good chance that no child will result.

Cloning Healthy Embryos

IVF is gruelling and distressing for couples who fail again and again to conceive. The emotional and financial cost of unsuccessful assisted conception can be high – so high that it may be better to give up, despite the heartbreak of childlessness. The creation of IVF embryos is easy enough; it is just that the overall success rate in getting an IVF pregnancy to term is relatively low, and we lack the techniques to correct genetic damage to improve that rate. Or rather, we lack a proven way to correct genetic damage accurately, confirm that it has worked, and only then transfer a modified human embryo to the womb. However, in the medium term nuclear transfer does offer a way to do this. Combined with genetic manipulation, it allows precise DNA changes to be made, including modification of existing genes. And it is possible to check in the laboratory that a genetic change has worked before a pregnancy is started. Since the birth of Polly, these methods have been used to create entirely new biomedical opportunities, as I discussed in Chapter 5.

The successful birth of Polly rested on the ability to grow cells in the laboratory for a long period during which genetic changes could be made. We used foetal cells to make her because they were available in sufficiently large numbers and have a long life in culture. However, we found that the procedure of genetic change, selecting those cells with the modification and confirming the change sometimes took too long, and the cells died or became abnormal in culture. This will not do for human medicine. For human medicine I envisage the use of embryonic stem cells, which can be grown almost indefinitely in culture, so that all the steps can be carried out. To prevent the birth of children with genetic disease, one would take an IVF embryo shown by pre-implantation diagnosis to have a hereditary defect, remove the stem cells, carry out the genetic correction, check that modification has worked, and then use nuclear transfer into an enucleated egg to create a new embryo without the disease.

The resulting child would be a clone not of any person but of an early embryo. For me, this distinction is profoundly important. As I explained before, I am extremely concerned about the effects on a child of being a clone of another person, and I oppose it. However, an early embryo is not a person, and I see the proposed use of nuclear transfer in this way to prevent a child's having a dreadful disease as far less controversial. One could even argue that this is not a true clone at all: although the genetic modification is slight, so that the cloned embryo's DNA differs by a fraction of a per cent from that of the early embryo that donated the nucleus, the resulting child would be spared an awful hereditary disease. Even to a layperson there would be a world of difference between these clones.

No doubt there would be intense media interest in the first birth by this method, and that could be stressful for all those involved in the pregnancy. But in my view this stress would be less than that of being a 'clone' – in the sense of being a genetically identical twin of another person – and would arise only in the first few early cases, before the technique became routine and was deemed no longer newsworthy by journalists. If it were safe and effective, I can see nothing immoral about this use of nuclear transfer to prevent disease.

These methods could be developed to introduce genetic changes into embryos in order to prevent the birth of children with cystic fibrosis, who die prematurely of lung disease, or with Huntington's disease, a devastating degenerative illness in which the patient loses control of their limbs to such an extent that they appear to dance. The original name for the disease, 'Huntington's chorea', means just that – 'Huntington's dance' ('chorea' comes from the Greek word for dancing). The first symptoms do not appear until middle age, but the decline is relentless, ending in death. Not surprisingly, young people at risk of this awful disease often reject the offer of a genetic test to show whether it will develop.

My proposal would not cure afflicted parents, but it would help them to ensure that any children they might have would be free of the disease. If an affected couple produced an embryo that carried the abnormal gene, revealed by a telltale 'genetic stutter' in the genetic code typical of Huntington's, then genetic modification of those cells before cloning to produce a new, stutter-free embryo would ensure that the child would not develop the disease. I strongly believe that, as soon as it becomes safe to do so, we should consider using cloning for this purpose. This method offers new hope to those who are unlucky and who consistently fail to produce healthy embryos. When I ask students what they think of using genetic correction and cloning this way, they point to the difficulty of doing it, scarcity of eggs, and so on. But they raise no fundamental ethical objections. I suspect – and I hope – that the public will feel the same way.

Enhancing the Germ Line

The child produced as a result of nuclear transfer using a genetically altered embryonic cell would grow to be an adult who could pass on this genetic change. This is germ line modification. A fertility treatment called cytoplasmic transfer has some of the same characteristics. In this experimental method the cytoplasm from a healthy donor egg is injected into the egg of an older woman to 'rejuvenate' it. The purpose of cytoplasmic transfer is ultimately to produce an embryo with the genetic make-up of both parents, providing an alternative to using donated eggs. But it may also have the effect of introducing a genetic change – foreign DNA in the mitochondrial power packs in the donor cytoplasm – that may be passed to future generations down the maternal line (we inherit only our mother's power packs).

The proposal is radical because scientists have agonised for at least three decades over what is called 'germ line therapy – genetic changes passed down generations in eggs and sperm – and remain divided over whether it should be allowed. On one side are those who would outlaw it because genes can interact in complicated ways and because the eradication of certain so-called disease genes could have unexpected side effects. They fear that errors could wreak havoc with a person's genetic legacy and pollute the human gene pool. Against them in the debate are the pragmatists who say that these are theoretical risks and should not be allowed to outweigh the very real potential of germ line therapy to reduce human suffering. But cytoplasmic transfer is not like the usual brand of germ line therapy – boys would not pass on the genetic change to their children. Since the first child to have been created this way, Emma, was born in 1997, a small number of others have followed. In a limited sense, they all have two mothers.

A number of prominent reproductive biologists, including Robert Winston, Alan Trounson and Roger Short, were not convinced that cytoplasmic transfer improved pregnancy rates. And although mitochondria are usually thought of as governing cellular energy production, rather than the phenotype (the outward physical appearance and constitution of a creature, influenced by DNA in the nucleus), there were concerns about unforeseen problems because mitochondria also play important roles in other cellular processes, such as cell death. The controversial work, which was stopped in 2001, was done at the Institute for Reproductive Medicine and Science of St Barnabas in Livingston, New Jersey, by a team that included a familiar figure in the cloning story, none other than Steen Willadsen.

The therapy I envisage would go much further and alter one in thousands of the instructions in the nuclear DNA, the DNA that shapes appearance, intellect and susceptibility to disease. Many scientists are convinced that we should never attempt this because the consequences are passed to future generations. But not everyone thinks that way. 'Whether based in ethics or unfounded fears of the unknown, such arguments are ultimately not compelling in my judgement,' asserts Jim Watson, who with Francis Crick and Maurice Wilkins helped to launch the molecular biology revolution by revealing the structure of DNA. 'Germ line therapy is in principle simply putting right what chance has put horribly wrong.' This is a view I share.

In his book *DNA: The Secret of Life*, Watson goes beyond the relatively uncontroversial application of fixing hereditary disease, to preventing disease in the first place: 'Germ line gene therapy has the potential for making humans resistant to the ravages of HIV. Some would say that, rather than altering people's genes, we should concentrate our efforts on treating those we can and impressing upon everyone else the dangers of promiscuous sex. But I find such a moralistic response to be profoundly *immoral*. Education has proven a powerful but hopelessly insufficient weapon in our war.'

Indeed, Watson goes further still, advocating genetic enhancement: genetic changes that address our desires rather than our needs.
This could lead to a social revolution, since we would then be
able – in theory at least – to mould our children in all kinds of ways.
In raising the possibility of viable germ line therapy, he says, 'Having
identified the relevant genes, would we want to exercise a future
power to transform slow learners into fast ones before they are even
born? We are not dealing in science fiction here: we can already give
mice better memories. Is there a reason why our goal shouldn't be
to do the same for humans?'

Despite the risks, Watson argues that we should give serious consideration to germ line therapy: 'I only hope that the many
biologists who share my opinion will stand tall in the debates to
come and not be intimidated by the inevitable criticism. Some of us
already know the pain of being tarred with the brush once reserved
for eugenicists. But that is ultimately a small price to pay to redress
genetic injustice. If such work be called eugenics, then I am a
eugenicist.'

One could argue that the blend of cloning and genetic modification can offer another route to enhancement, so that parents can
offer their children a 'better life'. In one sense, there is nothing new
in parents struggling to give their children a head start. They try to
get them into the best schools. They may sign them up for additional classes, whether in pottery, music or ballet. They ensure that
they are vaccinated. They give them antibiotics and warm them
with central heating to keep them well. They pay for expensive
dental work to correct their teeth. Some even sign their children up
for cosmetic surgery, or alter their mood with drugs such as Ritalin
and Prozac.

I suspect that many people would consider tinkering with the
genes of their children too. Proponents of enhancement would
accept that genes do not determine precisely who we are but
argue that they constrain an envelope of possibilities, and that

genetic alteration offers a way to increase the range of possibilities. For example, it might be used to alter genes of utility, such as those linked with memory or dexterity. It might alter genes valued by society, such as those that contribute to our capacity for altruism and sympathy. There are even wilder ideas, of altering humans so that they are more tolerant of pollution, or of climate change.

I have objections to the proposal to enhance the abilities of our children, on both biological and ethical grounds. Like most people, I disapprove strongly of the idea of an embryo coaxed to life for shallow reasons of status, preference or style. Any such work is unsavoury because it reduces children to consumer objects that can be 'accessorised' according to the parents' whims. As many have argued, love for one's offspring should not be contingent upon the characteristics they possess, even in our less than ideal world.

There is also a fundamental difference between genetic enhancement and established methods of enhancement, such as education, mood-altering drugs and cosmetic surgery. Genetic changes would be transmitted to all future generations, and their effects would be harder to reverse. They could undermine qualities and traits that are fundamental to our humanness, or exaggerate intolerance of disabilities. There is a need here for great caution.

There is also a major biological concern. I am sceptical that genetic enhancement is even possible because the genetic control of many traits is so complex. Any of the genes involved in any one trait also influence many other traits. In short, we could not predict the effect of changing genes except for inherited diseases associated with errors in a known gene. Discussions of 'designer babies' often turn to how genes control eye colour, which they do; and there may follow glib claims about creating blue-eyed, blond-haired babies. But even in this simple example the genetics is far from straightforward. The colour of our eyes is influenced by the inter-action of several genes that alter the distribution and content of pigment (melanin) producing cells, melanocytes, in the eye. Children are usually taught that brown is dominant, so if one parent has brown eyes their offspring can also have them, while two blue-eyed parents always have a blue-eyed child. But, though uncommon, blue-eyed parents can have children with brown eyes. Indeed, there are subtleties of eye colour and a range of hues that scientists cannot yet fully explain.

With such traits as intelligence, in addition to this genetic com-plexity, the environment – our family life, our education – of course has a considerable effect, and there is a limit to what can be achieved simply by snipping, inserting and shuffling DNA. Like most diseases, most traits will turn out to be polygenic – the product of an inter-play of genes with each other and with the environment. Intelligence, facial structure and extroversion are likely to be genet-ically complex. Even if it were possible to spot the crucial genes, enormous numbers of embryos would have to be tested to find the one with 'the right stuff'.

There is another problem facing parents who want to boost the brainpower of their child: genes can influence several characteristics. Any genetic change intended to influence intelligence, say, could also change other aspects of personality in an unpredictable way. One of the DNA-sequencing pioneers, Francis Collins, told a major conference in France a few years ago about some rich parents who hoped to produce a baby genetically enhanced to be artistic and musical but ended up with 'a sullen adolescent who smokes mari-juana and doesn't talk to them'. In short, each 'enhancement' would be an experiment. I find it unacceptable to experiment on children. I believe that we ought to accept ourselves and our children for what we are, rather than attempt to breed an improved race of human beings. Like all my peers, I think that money would be better spent on education than on genetic enhancement. Finding a good school will for some considerable time remain the best way to enhance a child's intelligence.

Others have raised conjecture about genetic engineering to heights that have no bearing upon reality. Lee M. Silver of Princeton University, the author of *Remaking Eden: Cloning and Beyond in a Brave New World*, raises the spectre of 'boutique babies': parents sit at a computer and scroll through a series of genetic menus from which they choose the characteristics they want for their offspring. He speculates how, in the distant future, the upper classes will have had turbocharged children with so much genetic enhancement over so many generations that they become unable to mate with members of other classes. The posh would be a different species to paupers, and he warns that this would be 'the most horrible thing that ever happened to humanity'.

I have confidence that posh and paupers will continue to exchange genes, whether by old-fashioned sex or by assisted reproduction, so that divergence into separate species will never happen. I appreciate that Silver is trying to make a vivid point, but this speculation would be more at home in a science-fiction novel. We have been here before, after all, with the beautiful Eloi and the brutish Morlocks of *The Time Machine*, the 1895 classic by H. G. Wells. However, I do accept Silver's argument that the driving force behind many uses of new technology is not governments, as in Aldous Huxley's *Brave New World*, with its state-run human hatcheries. Rather, it is commercial pressures. If we leave it to market forces, then I am sure that some couples will attempt to design their own children. If fictions like *Remaking Eden* or *Brave New World* teach us anything, it is that the unfettered use of scientific technologies can present a danger to human freedom and dignity. The way they are applied has to be regulated. As Huxley once said, science in itself is morally neutral; it becomes good or evil according to how it is applied.

Banning Designer Babies

Despite many people's almost religious belief in free markets, experience of the food and pharmaceutical industries shows us that an unregulated market is inappropriate for matters of public health and thus of genetic modification. In the United States, where the faith in market forces is strong, the production of food and medicines has been supervised for a century by the Food and Drug Administration.

At the beginning of the twentieth century there was little regulation of the sale of medicines. Many of the products that were sold had no clinical value. Those that did, like quinine-containing cinchona bark powder, could be made less effective – and much more profitable – by cutting them with just about anything, and others were laced with what we now class as recreational drugs. The latter may have reduced discomfort and perhaps made the patient 'happy' for a few hours, but they provided no treatment of the underlying condition.

Supervision of food production and marketing was also very limited. Upton Sinclair's influential novel *The Jungle* (1906) described the wretched sanitary and working conditions in the Chicago meat-packing industry and helped to bring about action. Publications such as the *Ladies' Home Journal* played a leading role in bringing political pressure to bear in Congress. A number of acts to tighten the regulation of production and marketing were passed. Despite these changes, there was still no obligation to assess the effect of new medicines, and tragedy was inevitable. A Tennessee drug company marketed a form of the new wonder drug, Elixir Sulfanilamide, an early antibiotic. However, the sweet-tasting solvent in this untested product was a highly toxic chemical analogue of antifreeze, and over a hundred people died, including many children. The public outcry led to Franklin D. Roosevelt's signing the Food, Drug, and Cosmetic Act on 25 June 1938. The thalidomide tragedy of the 1960s further expanded the FDA's authority. The moral of these grim stories is clear: it often takes a disaster to force governments to take action.

Past events warn us of the very real danger of tragic consequences if developments in reproductive technology escape regulation. In particular, great care must be taken if the entire future of a child is being determined. A regulatory framework is required. This is too important to leave to the scientists, the clinicians or even the prospective parents, all of whom have their own specific viewpoint. Rather, we should expect a framework to be prepared to guide those who are too closely involved, who may have vested interests or who may fail to appreciate the full consequences for the children who result. The framework could be enforced by national or regional regulation, or by a medical ethics committee. It is not really important which body provides the framework: the critical requirement is that guidance be provided by someone who is not directly involved.

Drawing the Line

Although I am for the use of genetic modification to treat disease and against enhancement, I would be the first to admit that there will be endless arguments over where to draw the line. Everyone understands the wish to keep children from being born with a serious genetic disease such as Huntington's, for which there is no treatment and whose consequences are devastating. If it becomes safe and effective to prevent this and other serious genetic disorders by cloning, it is hard to see how it could be objected to on ethical grounds. Similarly, most of us would say that it is desirable to administer synthetic growth hormone to ensure that a stunted child grows to normal height. By the same token, the use of germ line therapy to spare the children of that child a similar fate seems reasonable.

But what if a 'normal' child's parents cited the extensive and persuasive scientific literature on how tall people fare better in life – they tend to earn more, get more respect and attract partners more easily – and demand that the height of their normal child be boosted a little? Indeed, what if the parents wanted to guarantee that their child would be tall enough to be a basketball player?

People with an inadequate, abnormal or unusual physical appearance can become depressed and feel inadequate. Some parents might argue for cosmetic changes on the grounds that they could save a life by preventing vulnerable young people from becoming suicidal about the size of their nose or their baldness, say. The risks that people take with cosmetic surgery are a powerful testament to the desire for a body image that conforms to a perceived norm, and there will doubtless be intense pressure from some parents to carry out genetic surgery on their unborn child to achieve similar ends.

The possibilities of genetic tinkering raise other issues. Is ageing a disease? Surely not – it is a normal part of the life cycle. But, given recent research that has extended the life of creatures such as the nematode worm, it seems highly likely that there are human genes that influence longevity (though they are by no means the whole story when it comes to a long life). Does that mean that old age should be 'cured' by genetic modification? And if so, what does that mean for society? John Harris, professor of bioethics at Manchester University, has suggested that this 'creeping longevity' would have profound implications for what we mean by the sanctity of human life, increasing resentment as the super-old competed with the young for jobs, space and other resources. He suggests that society may even resort to what he calls 'generational cleansing': once individuals have had a 'fair innings', they must die – by suicide, euthanasia or reactivation of the ageing process.

Whatever the shade of grey between enhancement and therapy, and whatever boundary is being transgressed, one aspect of this medical intervention stays the same: genetic enhancement by the method I have outlined in this chapter *will* be proposed by parents on behalf of their unborn child. This is about the fate of children of the future, a decision that has only indirect impacts on the person making the choice. Society has an obligation to intervene on the embryo's behalf when it comes to weighing the risks and benefits of genetic alteration. Selection for traits thought by the parents to be beneficial could later be seen by the child as a curse. Parents striving to achieve their goals and make their investment worthwhile may discover that the uninhibited selection of children may erode the unconditional love that is the bedrock of the parent–child relationship.

Apart from these fundamental moral and ethical issues, there are practical and social considerations. Genetic correction will be expensive. Several complex and difficult steps are required, and at each point careful checks must be made to ensure that all is well. To make this process reliable will take time and cost money. This means that choices will have to be made about how to allocate finite resources. As ever, the benefits are most likely to be enjoyed by the rich first, suggesting there will be genetic haves and have-nots. Hospitals and health providers are already well versed in these issues. Use of this method could also be curbed by the limited availability of eggs, for which there are other demands, notably in the treatment of infertility. In short, the genetic modification of the human gene pool will have to be considered in a broad context, that of society as well as the family. And perhaps of history too.

In many of the discussions of where my research is taking us, one senses that the benchmark used to judge whether reproductive science is helpful or harmful is a utopian zero-risk society, a society free of narcissism, a society where children are much more than chattels, and a just society free of the taint of self-interest. But over the millennia, pressure from family, peers and society has thrown together couples to bind allegiances, achieve status and combine wealth. Children have been conceived to perpetuate businesses, estates and royal bloodlines. In some ways, 'designer babies' are nothing new: think of the mother with three daughters who is desperate to conceive again to have a son; the career woman approaching menopause who feels she had better conceive quickly so that she can claim to have led a 'full life'; the poverty-stricken who have a big family to help work the farm; and the celebrity who thinks that a baby in designer denim is this year's must-have accessory.

The effects of these traditional splashings in the human gene pool, not a utopian ideal, should provide the upper benchmark for what should and should not be allowed when we apply reproductive science: we should aspire to do much better than this and hope that we never do any worse. Whatever we do, I doubt that this science will cause more than ripples when compared with the waves of evolutionary change that came before.

Envoi

Beyond Human Cloning

The spin, hype and exaggeration that have so far greeted my work suggest a dizzying pace of progress in research on assisted reproduction carried out by wild-eyed Frankensteins, with science out of control and accelerating far too quickly for irritating little things like ethical considerations to have the smallest hope of keeping up. The truth is very different. We have plenty of time to sort out the ethics, because in comparison with the rates of take-up in other fields of science, the application of reproductive technology is slow and painstaking, and is already highly regulated in many countries.

Take the development of *in vitro* fertilisation, for example. Having begun in earnest in 1962, IVF research has passed many technical milestones, from egg maturation and fertilisation in the test tube to methods of stimulating the growth of follicles. As if to underline the revolutionary implications of what they were trying to do, Robert Edwards and Patrick Steptoe were denied government funding and attacked by the 'Holy Trinity': the pope, the press and prominent Nobel laureates. (Jim Watson demanded of Edwards at a 1971 press conference, 'What are we going to do with the mistakes?')

IVF embryo transfers began only in 1972, a decade after this IVF research began, and were rewarded by a first clinical pregnancy in 1976, unfortunately ectopic, and then the birth of Louise Brown in 1978. The technique has since been shown to be as safe as one could expect, and has brought joy to millions of people around the world. Debate continues, of course. Since the birth of Louise there have been arguments about whether IVF should be used to enable postmenopausal women to have children, what the fate of spare and frozen embryos should be, and whether every woman has the right to have a child. But we have had many years to mull over the implications. This long debate and the obvious benefits for the infertile have built public confidence in IVF.

A similar history of early failure and social controversy followed by eventual technical success and public acceptance could be charted for many biological innovations, including organ transplants and the introduction of anaesthetic during childbirth. Ambition and adventure vary from country to country, but the general pattern is similar everywhere.

We are also fortunate that researchers have been willing to enter into discussions about the implications of their work. I hope with this book to continue a long tradition of scientists being open about what they do and in speculating on where their research is taking us. In the case of germ line modification, a paper published in the journal *Science* in 1972 shows that scientists were thinking of the consequences even then. And they continue to do so, for instance in a recent report by the American Association for the Advancement of Science. Moreover, we have active media that show a fertile but sometimes overactive imagination when it comes to revealing the future path and pitfalls of a particular technology.

But I have faith in more than openness, debate and regulation. I have faith in people too. I have confidence that a well-informed democracy can keep abuses in check; I have confidence in women who would not donate eggs to clone a dictator, but would to help a patient with an urgent clinical need; and above all else I have faith in the vast majority of scientists, who are no different from anyone else in wanting to reduce suffering and make the world a better place.

There is one crucial caveat I must add. No matter how powerful science becomes, the future will remain mysterious and open. You can predict neither the outcome of research nor the uses to which new knowledge will be put. The birth of Dolly opened the door to a new realm of medicine and biology. After Dolly we can glimpse some of the treasures and perils in this realm, but we still have much more to learn.

Just as the influence of the Internet today could never have been foreseen when Tim Berners-Lee invented the World Wide Web in 1989, so we should also expect methods of cloning to be unrecognisable twenty years from now. As for what will be possible in a century's time, that is anybody's guess, though I imagine that someone out there, whether scientist, futurist or science-fiction writer, has already thought about it to some extent.

Whatever emerges, I am confident that the technology will become safer, that there will be embryo-free alternatives to the use of nuclear transfer in the creation of stem cells, and that its combination with genetic modification will become routine. Most important, I am sure that many of the difficult ethical issues will be thrashed out to the satisfaction of the majority of society.

It is crucial that we do not allow our exuberant creativity to be curbed by a fear that new knowledge could be misused. We often take for granted what has been hard-won from previous research, and overlook the possibilities and potential of future developments that may well dwarf the attainments of the past. As much damage can be done by failing to exploit the beneficial applications of a technology as by promoting applications of that technology which are risky or harmful.

Assisted reproduction technology is very much in its Stone Age, and the revulsion that my proposal will stir in some people says less about the technology itself and more about the shock of the new, as it once did in the early days of heart transplants, vaccines and other innovations. History – the future-that-was that came in the wake of these early developments – shows how utility can breed acceptance.

The case for the genetic alteration of humans as I have outlined it in this book will always have its critics. Even when the technologies of nuclear transfer, genetic manipulation and stem cells have matured, I am sure that some people will still prefer to put up with the random insults of nature than be subject to human intervention, even if it is based on careful consideration of medical issues rather than a whim. They are, of course, free to turn their backs on the benefits of new technology. But at least they will have a choice. And for me, just having the chance to decide is paramount.

Although I have no idea what the future will bring, I am confident that reproductive and genetic technologies will greatly expand our possibilities. By the same token, they will also expand the burden of responsibility. Because of my faith that the majority of people know right from wrong, I feel that the sooner we take on that burden the better.

I want to be able to change my destiny rather than be condemned to a particular fate. I want people to have new options when it comes to that most fundamental urge to bring healthy children into this world. For me, the widespread use of genetic and reproductive technologies is not a step backwards into darkness, but a step forwards into the light.

Glossary

Terms in **bold** are explained elsewhere in the glossary.

activation The process during fertilisation by which the sperm induces the egg (**oocyte**) to resume its development. This triggers the completion of the second phase of **meiosis**, the process by which the egg halves its number of **chromosomes**. The activation caused by the sperm can be mimicked by treating the egg with selected chemicals or applying an electric shock.

Alzheimer's disease The most common form of dementia. Protein clumps called 'plaques' and 'tangles' develop in the brain, leading to the death of cells. As more cells die the symptoms become more severe, from memory lapses and confusion to mood swings, with death following eventually. In some families, this neurodegenerative disease is inherited.

amino acid One of the molecular building blocks of proteins.

asexual reproduction Reproduction that does not require the union of egg and sperm and in which all the genetic material of an offspring comes from a single progenitor. In other words, cloning.

base The molecular units – 'letters' of genetic code, called nucleotides – that provide the variation in a strand of **DNA** or **RNA**. DNA has four different types of base – thymine, cytosine, adenine and guanine – while RNA contains uracil instead of thymine. The two molecular 'letters' on opposite – complementary – DNA or RNA strands are called a *base pair*.

blastocyst A fertilised egg when it reaches the stage of a microscopic ball of about forty to two hundred cells, sometimes referred to as a 'pre-embryo'. The cells form a hollow sphere, comprising an inner cell mass of embryonic stem cells that will become the foetus surrounded by an outer ring of cells – the trophectoderm – that will become part of the placenta.

cell The building block of bodies. A discrete, membrane-bound portion of living matter, the smallest unit capable of an independent existence. Complex organisms such as humans are made up of more than two hundred different types, such as muscle, bone and nerve. Overall, there are two basic types: somatic (body) cells and germ line (reproductive) cells – eggs and sperm. Forty human cells placed in a line would stretch about one millimetre.

cell cycle The life story of a cell, which can be divided into a period of growth during which the chromosomal **DNA** is copied and cell division (by **mitosis** or **meiosis**) occurs. The story consists of various chapters: the phase when DNA is replicated (S phase), the phase when the cell actually divides into two cells (M phase), the two intervening gap phases (G1 and G2) and a non-dividing state called **quiescence** (G0).

cell division During normal human cell division, the **chromosomes** align neatly in the middle of the cell into twenty-three pencil-shaped pairs of parallel strands called **chromatids**. At the same time, two **centrioles** form at opposite poles of the cell. Threadlike **microtubules** attach each of the chromatids to the centrioles, forming what is called the **spindle**. When the cell divides, the chromatids separate and are pulled by the microtubules to opposite poles. Thus, two identical pools of chromatids, the inherited genetic instructions of the future daughter cells, form at the poles. The interior of the cell separates in dumbbell-like fashion with a narrow bridge that eventually pinches in two to form the two daughter cells.

centriole A cellular **organelle** near the cell's nucleus consisting of a cylinder with nine microtubules arranged peripherally in a circle. Centrioles help to form the **spindle** during cell division.

chimera An embryo or offspring made up of cells from more than one embryo. The resulting animals are blends of cells from what would normally be two separate animals, each with its unique **DNA**.

chromatid One of two identical strands into which a chromosome splits during **mitosis**, or cell division.

chromatin That portion of the cell's **nucleus** that contains all of the **DNA** of the nucleus in animal or plant cells. When cells divide, the chromatin is seen as distinct **chromosomes**, duplicating, with an equal partition of each set of chromosomes then travelling to each of the new daughter cells.

chromosome A carrier of genetic information in the form of **DNA** that has been tightly wrapped up, with specialised proteins such as **histones**, into a package. This package is the visible form that DNA takes within the cell. Usually an animal's DNA is stored in many chromosomes, like volumes of an encyclopedia. In humans, all cells other than eggs and sperm usually contain forty-six chromosomes: twenty-two pairs of autosomes (chromosomes carried by both men and women) and either a pair of X chromosomes (in women) or an X chromosome and a Y chromosome (in men). In each pair of autosomes, one chromosome is inherited from an individual's father and one from the mother.

cleavage The process of cell division that allows the **zygote** (the newly fertilised egg) to develop into a **blastocyst**.

clone Clones are genetically identical individuals produced by a form of **asexual reproduction**. Plant cuttings are clones. In this book, 'clone' is used to describe offspring produced by transfer of the nucleus from one cell into an egg (**oocyte**) from which the genetic information has been removed: the process of **nuclear transfer**.

cytoplasm The material in a cell that surrounds the inner sanctum, called the **nucleus** (where the **DNA** resides), and is bound by the cell wall. The cytoplasm is highly organised and contains specialised structures called **organelles**.

cytoskeleton The network of fibres that form the 'skeleton' of a cell.

culture The feeding of cells to keep them alive in the laboratory. Typically, cells and embryos are grown in a special liquid, known as a culture medium, which provides all the energy and other resources that are required for the well-being of the cells.

diabetes A disorder of the metabolism that strikes when the body cannot balance the storage and use of energy because of an error in either the production of the hormone **insulin** or the response to insulin. There are two main types of diabetes. In *type I diabetes*, also called insulin-dependent diabetes, the patient does not produce adequate amounts of insulin because insulin-producing cells in the pancreas – **islet cells** – have been destroyed by the body's immune system. Such patients respond well to injections of insulin. In *type II diabetes* the person no longer makes enough insulin or no longer responds to injected insulin. Type II is the more common form; it usually occurs in adults and is linked to obesity.

differentiation The process by which a primitive stem cell from an embryo 'commits' to becoming a specialised cell in the body, such as a skin cell or a bone cell.

diploid Describing a cell that contains two sets of chromosomes, as opposed to a **haploid** cell, which contains only one set. A diploid human cell contains forty-six chromosomes; a haploid – egg or sperm, for example – contains twenty-three.

DNA Deoxyribonucleic acid, the vehicle of inheritance for all creatures.

The complex giant nucleic acid molecule carries the genetic recipe for the design and assembly of proteins, the basic building blocks of life. The chain-like DNA molecule is in the form of a double helix made up of a series of **bases** which come in four types (adenine, guanine, cytosine and thymine, or A, G, C and T). The order of these bases provides a recipe (first issued through the medium of another molecule, **RNA**) for the proteins that are the building blocks of life. The code is a three-letter one, with a triplet of letters (ATT, say) coding for a particular amino acid that, when joined with a string of others, makes a protein. Each person has about three billion bases in his genetic make-up, or genome, but only around 25,000 genes that work to make proteins.

dopamine A substance found in the brain that transmits messages between nerves. The brain disorder **Parkinson's disease** is a result of inadequate functioning of dopamine-producing **neurons** in a part of the brain called the substantia nigra. Dopamine allows smooth, coordinated function of the body's muscles and movement. When approximately 80 per cent of the dopamine-producing cells are damaged, symptoms of Parkinson's disease such as tremor and slowness of movement appear.

Down's syndrome A genetic disorder that is caused by the presence of an extra copy of chromosome 21, or a piece of one, leading to a range of physical and mental effects in varying degrees. Physical characteristics include a flattened face, more widely spaced and slanted eyes, and smaller head size. Mental retardation is also typical, though there are wide variations in ability and behaviour. Possible related health problems include poorer resistance to infection, hearing loss, gastrointestinal problems and heart defects.

ectoderm The part of a developing embryo that gives rise to epidermis, the outer layer of the skin, and nerve cells. It is one of three germ layers – the basic three cellular layers, ectoderm, **endoderm** and **mesoderm** – from which the organs and tissues of the body develop through further differentiation.

embryo An organism in the early stages of growth and differentiation. During sexual reproduction it is formed at fertilisation by the union of sperm and egg.

embryonic stem cell A **stem cell** in the early embryo. Stem cells have the characteristic of being able to divide many times and so give rise to the two hundred or so cell types in a body. Embryonic stem cells retain the ability to form any of the different tissues of an adult.

endoderm The part of an embryo that gives rise to the epithelium (lining) of the gut and gut derivatives. In vertebrates, such as sheep and humans, these derivatives include the liver, trachea and lungs. The endoderm is one of three germ layers. See also **ectoderm**.

enucleate To remove a nucleus from a cell, usually an egg (**oocyte**), in a process carried out during **nuclear transfer**. The term is somewhat misleading: the egg cells often used in cloning are in a special stage called MII oocytes and don't actually have a nucleus, but free chromosomes instead.

enzyme A biological catalyst usually composed of a large protein molecule consisting of many, even thousands, of atoms that accelerates essential chemical reactions in living cells. (The name is from the Greek for 'in yeast'.) While a cell has a size measured in micrometres (millionths of a metre), enzymes and other proteins are of nanometre size (a billionth of a metre).

epigenetics Changes in how genes are used in the body that take place without changes to the genes themselves, notably during development. It is these epigenetic changes that make, say, a lung cell so distinctive from a brain cell, even though they contain the same **DNA**. One way to achieve epigenetic modification is to alter the way DNA is decorated with methyl groups, a process called DNA methylation. Another is through the action of **histones**. In epigenetic reprogramming, genes that were turned off during embryonic development and cell differentiation can become active again. In other words, the developmental clock is wound back so that the epigenetic modifications that turned an embryonic stem cell into a liver cell, for example, are removed. With the developmental slate wiped clean, this cell can once again be persuaded to develop into various types.

eugenics An attempt to alter (with the aim of 'improving') the genetic constitution of future generations by controlling reproduction. It was popular in the first part of the twentieth century, when individuals with traits considered to be desirable were encouraged to reproduce, and those with traits considered undesirable were discouraged from having children. Thousands were involuntarily sterilised. The ideas of the eugenics movement were models for the Nazis, whose eugenic ideologies culminated in the Holocaust.

evolution The idea of shared descent of all creatures, men, sheep and mice included. The word is from the Latin *evolutio*, meaning 'unfolding'. The names of Charles Robert Darwin (1809–82) and Alfred Russel Wallace (1823–1913) have long been joined with the modern concept of evolution and the theory of natural selection. In his *Origin of Species* (1859), Darwin suggested a mechanism for evolution. Natural selection is the process by which biological individuals that are endowed with favourable or deleterious traits end up reproducing more than or less than other individuals that do not possess such traits. In time, Darwin suggested, this led to the evolution of new forms of living organism – the origin of species. His ideas fit perfectly with those of modern genetics. Diversity arises through mutation, random changes in the genetic material **DNA**, from generation to generation.

fertilisation The fusion of sperm with egg, during which the sperm contributes genetic material to the egg (**oocyte**) and stimulates activation of that oocyte.

fibroblast A cell from connective tissue that is commonly grown in the laboratory because of the ease with which it can be cultured and made to proliferate.

gastrula A phase of embryonic development in which the three germ layers, the **endoderm**, **ectoderm** and **mesoderm**, have formed.

gene The unit of heredity, consisting of a **DNA** sequence that directs production of a particular protein and determines the tissues in which it will be made and at which the stages of development. Everyone inherits two copies of each gene, one from each parent. A dominantly inherited genetic disease such as **Huntington's disease** occurs when only one copy of the gene is sufficient to produce the disease. A recessively inherited disease occurs only if both copies of the defective gene are present, as happens with Tay–Sachs disease, cystic fibrosis and sickle-cell disease.

gene therapy The transfer of DNA – either packed in **viruses** or in fat particles or even as naked DNA – to treat a genetic disorder. Clinical trials of gene therapy have been under way since 1990. Although results have been disappointing so far, there are several cases where it has been shown to work. The hope is that it will eventually be used to treat many diseases.

genetic code The sequence of chemical building blocks of **DNA** (bases) that spells out the instructions to make amino acids, the molecular building blocks of **proteins**.

genetic engineering Tinkering with an organism's **genetic code** – in the form of DNA – to produce animals and plants with desirable properties.

genome The entire **DNA** sequence of an organism, consisting of a series of 'letters' of genetic code (A, C, G and T). The human genome is around three billion letters long.

genomics The science of identifying the sequence of DNA – that is, the **genome** – in various species, and subsequent processing of that information.

germ cells Cells that give rise to egg or sperm.

germ line The lineage of cells that connect our generation to the previous one and even to the first generation of life four billion years ago. The sperm and the egg are examples of germ line cells.

haploid Describing a cell, such as an egg or sperm, that contains only twenty-three chromosomes, half the number in the other (**diploid**) cells in the body.

HFEA Human Fertilisation and Embryology Authority, a UK body that regulates procedures in which human embryos are used for research or therapy.

histone A protein that can bind to **DNA** and wrap the genetic material into 'beads on a string', in which DNA (the string) is wrapped around small blobs of histones (the beads) at regular intervals. Histones play a role in **chromosome** architecture and influence the way in which genes are used in the body.

hormone A substance produced by one tissue and conveyed by the bloodstream to another for metabolism and growth.

Huntington's disease A devastating inherited disorder marked by abnormal body movements, called chorea, and loss of memory.

implantation The process in which an embryo, after travelling through the oviduct (fallopian tube) to the uterus, embeds itself in the lining of the **uterus** so that development can continue.

imprinted gene A family of genes that are used only if they are inherited from the mother, or from the father. We have two copies of most genes; exceptions include the genes that determine gender in males. Usually the two copies are used equally in the body, but with imprinted genes only the copy inherited from one parent is used. **Epigenetic** mechanisms control imprinting.

inner cell mass A clump of cells growing within and to one side of the **blastocyst**, from which the embryo develops.

insulin A protein **hormone**, released into the circulation by the pancreas, that regulates the concentration of glucose in the blood. **Diabetes** is caused either by a lack of insulin or by lack of a response to insulin.

in vitro Performed outside the body, that is in the laboratory (the literal meaning of *in vitro* is 'in glass', for example in a test tube).

islet cell A pancreatic cell that produces hormones, such as **insulin**, which are secreted into the bloodstream. These hormones help to control the level of glucose (sugar) in the blood.

IVF *In vitro* fertilisation, the union of an egg and sperm outside the body and in an artificial environment such as a test tube or, more commonly, in a Petri dish.

large–offspring syndrome A condition often seen in cloned cattle and sheep in which foetuses grow much larger than their normal size in the

uterus and often suffer organ defects. The syndrome poses serious health risks to the mother and to the offspring, often resulting in the death of one or both.

meiosis The formation of **germ cells** – eggs and sperm – where the chromosome number is cut by half. At the end of this specialised form of cell division, which can take decades in the case of eggs (oocytes), the new cells have only half the number of chromosomes of the source cells.

mesoderm The part of the embryo that gives rise to the muscular and circulatory systems and most of the skeletal and urogenital systems. The mesoderm is one of three germ layers. See also **ectoderm** and **endoderm**.

metaphase The phase of cell division in which all the chromosomes are assembled by the **cytoskeleton** before **cleavage**.

microtubule A component of the **cytoskeleton**. Microtubules have a particular role in separating the pairs of **chromosomes** during cell division. In the egg at metaphase II of meiosis, the chromosomes are held on a **spindle** ready for division.

mitochondrion A small energy-producing structure, a lozenge-like capsule that is one of the **organelles** found inside cells. In most cells there are several thousand mitochondria. The egg (**oocyte**) at ovulation has a particularly large number, but no new mitochondria are produced before the **blastocyst** stage. The **DNA** of all mitochondria in a body, whether male or female, is obtained from the mitochondria in the egg, so this genetic code is passed down the maternal line only. Mitochondria give rise to other mitochondria by copying their small piece of mitochondrial DNA and passing one copy of the DNA to each of the two resulting mitochondria. In the great majority of cases, the mitochondria of cloned offspring are all derived from the egg, not the donor of the DNA.

mitosis The process of cell division where the **DNA** is replicated and one cell becomes two. Mechanisms in the mitotic cell cycle ensure that the number of chromosomes in each daughter cell is the same as in the parent cell.

molecular biology The study of the molecular basis of life, including the biochemistry of molecules such as **DNA** and **RNA**.

morula The embryo at the stage when it is a ball of closely-knit cells, before it becomes a **blastocyst**.

mutation A change in genes produced by chance or deliberate change in the **DNA** that makes up the hereditary recipe of an organism.

motor neuron disease A group of disorders in which motor nerve cells (**neurons**) in the spinal cord and brain stem deteriorate and die.

Amyotrophic lateral sclerosis (ALS), also called Lou Gehrig's disease, is the most common motor neuron disease.

MPF Maturation promoting factor (or meiosis or mitosis promoting factor), a complex protein made up of several protein subunits that acts at a specific point in the cell cycle to cause the breakdown of the envelope that surrounds the nucleus and the packaging of DNA into **chromosomes**, so-called chromosome condensation. MPF plays an important role during nuclear transfer.

neuron The nerve cell that is the fundamental unit of the nervous system.

nucleic acid A complex organic acid made up of a long chain of units called bases. The two types, known as **DNA** and **RNA**, form the basis of heredity.

nuclear transfer The most common form of cloning, in which a nucleus is transferred from a donor cell into an egg (**oocyte**) from which the chromosomes have been removed. The genetic information in that nucleus determines almost all the characteristics of the resulting off-spring.

nucleus The small region in the centre of a cell that contains the DNA/chromosomes – the instructions to build and run the cell.

oocyte A term used loosely to describe the egg at several different stages of **meiosis**. At the stage when oocytes are shed from the ovary, they are at the second metaphase of meiosis (MII), which is the preferred stage for use in **nuclear transfer**.

organelle Any of various structures found within cells that are analogous to organs in a body. Organelles are too small to be seen with an optical microscope.

ovary The female reproductive organ that produces egg cells and the steroid hormones that govern many aspects of reproductive behaviour and the function of other organs of the reproductive tract, such as the uterus.

ovulation The release of eggs from the ovary.

Parkinson's disease A progressive neurodegenerative disorder caused by damaged or dead **dopamine** neurons in a region of the brain that controls balance and coordinates muscle movement.

parthenogenesis The development of embryos from unfertilised eggs that are activated by stimulation with electrical pulses or chemicals. (The word means literally 'virgin birth'.) Such embryos are called *parthenotes*.

placenta The organ that is formed during pregnancy to nourish the developing foetus. It is made up of tissue from both the foetus and the mother.

pluripotent Describing cells able to develop into cells of various different types, but not all possible types. Only **totipotent** cells can form all cell types.

polar body The means by which a cell sheds **chromosomes**. The body is a small structure outside the egg (**oocyte**) that contains the chromosomes that were discarded during **meiosis**. During experiments its presence is used to indicate the stage of meiosis that the oocyte has reached and to provide an indication of the probable location of the chromosomes that have to be removed.

primitive streak A thickening in surface of the embryo that marks the first clearly recognisable structure that will go on to form the nervous system. It is used by many scientists as a benchmark to help separate an early embryo from a 'person'.

pronucleus A small round structure seen within the egg after fertilisation. Pronuclei contain the **haploid** sets of **chromosomes** (genetic material of egg and sperm) surrounded by a membrane. A normal fertilised egg should contain two pronuclei, one from the egg and one from the sperm.

protein A large molecule composed of one or more chains of amino acids in a specific order, determined by the gene coding for the protein. Proteins are required for the structure, function and regulation of the body's cells, tissues and organs. Examples are hormones, enzymes and antibodies.

quiescence When a cell is comparatively inactive. Some cells in the body are quiescent while they await an instruction from the body to carry out a specific task, but quiescence can be induced in cells in culture either by starving them or by allowing them to become overcrowded.

reprogram Changing the repertoire of genes used by a cell. At any particular time the nucleus of a cell is organised to direct the production of a range of proteins that is appropriate for its function. A heart cell, for example, would use a spectrum of proteins different from that of a liver cell. The nucleus is said to be reprogrammed if the repertoire of genes that is in use is altered so that the nature of the cell is changed. The nucleus is also reprogrammed by transfer into an **oocyte**, but this process is inefficient in present procedures, leading to the occurrence of abnormalities of development.

RNA Ribonucleic acid, one of the two types of genetic material – in the form of nucleic acids – found in all cells. The other is deoxyribonucleic acid (**DNA**). RNA transmits genetic information from DNA to direct the proteins produced by the cell.

RNAi RNA interference, a way to turn off genes by introducing double-stranded RNA into a cell.

siRNA Small interfering RNA (siRNA), also known as short interfering RNA, is a class of RNA molecules that interfere with the use of genes. They are naturally produced as part of the RNA interference (**RNAi**) pathway.

spindle Before a normal cell divides, its chromosomes are duplicated and then pulled apart by a structure called a spindle, so that the two daughter cells will each have the same number of **chromosomes**. At the end of a normal spindle is the *spindle pole*, also called the centrosome, which pulls the chromosomes outwards.

stem cell The progenitors of the many different types of cell in the body. In most cases, if a cell divides it is able to give rise only to daughters that are like itself. In addition, such a differentiated cell is able to divide only a limited number of times. By contrast, stem cells have a greater ability to divide and give rise to other cell types. Adult stem cells give rise to tissues, and efforts are under way to see how flexible they are.

therapeutic cloning The production of cells from cloned human embryos either for use in research or to treat a disease. The stem cells are obtained from the **inner cell mass** of the embryo and then, by the use of growth factors, persuaded to develop into the required type.

totipotency Totipotent cells are able to give rise to cells of all the different tissues of an adult. By contrast, **pluripotent** cells can give rise only to a limited variety.

transdifferentiation The process by which a non-stem cell transforms into a different type of cell, or by which an already differentiated **stem cell** creates cells beyond its usual repertoire. For example, the direct conversion of a nerve cell to a muscle cell takes place by transdifferentiation.

transgenic An organism that has a foreign gene (a transgene) incorporated into its **genome**.

twinning Also called blastocyst splitting; the natural form of cloning, occurring when a **blastocyst** divides in two.

uterus The hollow pear-shaped organ in the lower abdomen of a woman in which the foetus develops. Also known as the womb.

virus One of the smallest infectious agents, consisting of a piece of genetic code wrapped in protein, measuring between 15 and 300 nanometres across. Viruses are responsible for a huge range of diseases, such as influenza. It is debatable whether viruses are living, since they have to hijack the molecular machinery of our cells to reproduce (they do this by **reprogramming** our cells with their genetic code, turning those cells into virus factories). They can themselves be hijacked to carry out **gene therapy**.

xenotransplantation The transfer of cells or tissues from one species to another.

zona pellucida The translucent membrane that encloses the eggs (**oocytes**) and early embryos of mammals. The embryo 'hatches' from the zona at the **blastocyst** stage of development.

zygote The one-cell embryo formed at fertilisation by the fusion of sperm and egg (**oocyte**).

Bibliography

Chapter 1: A Star Is Born

Interviews and Correspondence
Campbell, Keith. Phone interview with Roger Highfield, 16 June 2005.
Campbell, Keith. E-mail correspondence with Roger Highfield, 4 and 5 July 2005.
Kendrick, Keith. E-mail correspondence with Roger Highfield, 23 August 2004; 26 April and 12 August 2005.
King, Tim. E-mail correspondence with Roger Highfield, 6 April and 16 June 2005.
King, Tim. Interview with Roger Highfield, Roslin Institute, 14 June 2005.
Miller, Dusty. E-mail correspondence with Roger Highfield, 11 and 14 April 2005.
Rhind, Susan. Phone interview with Roger Highfield, 28 March 2005.
Weldon, Fay. Conversations with Roger Highfield, The Berkeley Hotel, London, 3 November 1994; The Savoy, London, 21 November 2002, 24 February 2004; The Groucho Club, London, 26 September 2003; Joint visit to the Roslin Institute, 18 November 2003.

Journal and Newspaper Articles
Cohen, Philip. 'Clone alone'. *New Scientist*, 9 May 1998 (no. 2133), p. 3232.
Dow, Ian. 'I'm not God: Prof is set to clone human cells'. *Daily Record*, 9 February 2005, p. 6.
Dyer, Geoff and David Firn. 'Death of Dolly the sheep fuels fears about cloning'. *Financial Times*, 15 February 2003, p. 3.
Grice, Elizabeth. 'My long search for liberty'. *Daily Telegraph*, 19 January 1998, p. 13.
Highfield, Roger. 'British scientists condemn plans to clone babies'. *Daily Telegraph*, 10 March 2001, p. 8.

Highfield, Roger. 'Scientists boycott "human clone" conference'. *Daily Telegraph*, 12 September 2001, p. 11.

Highfield, Roger. 'Fertility experts expel human clone doctor'. *Daily Telegraph*, 15 September 2001, p. 18.

Highfield, Roger. 'Human clone claim is met with scepticism'. *Daily Telegraph*, 28 November 2002, p. 9.

Highfield, Roger. 'Rival joins race to clone first human'. *Daily Telegraph*, 3 December 2002, p. 8.

Highfield, Roger. 'We've cloned baby, says cult'. *Daily Telegraph*, 28 December 2002, p. 1.

Highfield, Roger. 'From science fiction to unproved fact'. *Daily Telegraph*, 12 August 2004, p. 4.

Jaenisch, Rudolf and Ian Wilmut. 'Don't clone humans!' *Science* 30 March 2001 (no. 291), p. 2552.

Kolata, Gina. 'Some scientists ask: How do we know Dolly is a clone?'. *New York Times*, 29 July 1997, p. C3.

Mestel, Rosie. 'Dolly's demise puts cloning into question'. *Orlando Sentinel*, 23 February 2003, p. G6.

Milne, Shaun. 'Dolly makes opera debut at Baa-rbican'. The *Mirror*, 17 September 2002, p. 10.

Nesmith, Jeff. 'Scientists disagree on how to curb human cloning'. Wire report, Cox News Service, 25 June 1997.

Pennisi, Elizabeth. 'Cloning: The lamb that roared'. *Science* 19 December 1997 (no. 278), pp. 2038–9.

Reuters. 'Scientists clone the unexpected'. *Chicago Tribune*, 25 February 1997, p. 1.

Vendantam, Shankar. 'British scientist who cloned sheep urges caution at Senate hearing'. Wire report, Knight Ridder/Tribune News Service, 12 March 1997.

Wade, Nicholas. 'Scientist who announced cloning of sheep Dolly is taking steps to win over critics'. *New York Times*, 28 February 1998, p. A6.

Watson, J. D. and F. H. C. Crick. 'A structure for deoxyribose nucleic acid'. *Nature* 171 (25 April 1953): 737–8.

Wooton, Sarah K., Christine L. Halbert and A. Dusty Miller. 'Sheep retrovirus structural protein induces lung tumours'. *Nature* 434 (14 April 2005): 904–7.

Books

Kolata, Gina. *Clone: The Road to Dolly and the Path Ahead*. London: Allen Lane, 1997.

Watson, J. D. *The Double Helix*. Edited by G. Stent. London: Weidenfeld & Nicolson, 1981.

Wilmut, I., K. Campbell and C. Tudge. *The Second Creation*. London: Headline, 2000.

Internet Sources
Genetics Policy Institute <http://www.genpol.org>.
National Right to Life <http://www.nrlc.org>.
Reich, S. and B. Korot. 'Three tales'. The Steve Reich website <http://www.stevereich.com>.
The Raelian Movement presents: Message from the Designers <http://www.rael.org>.
The Zavos Organization <http://www.zavos.org>.

Stage Play
Churchill, Caryl. *A Number*. Royal Court Theatre, London, 23 September–16 November 2002.

Chapter 2: Cloning the Cloner

Interviews and Speeches
Emlen, Stephen T. 'Sexual differences in mating strategies: The interface of culture law and biology'. Speech given to the annual meeting of the American Association for the Advancement of Science, Seattle, Washington, 14 February 1997.
Evans, Martin. Interview with Roger Highfield, Biochemical Society meeting, Nottingham, 20 July 1988.
Evans, Martin. Phone interview with Roger Highfield, 9 March 2005.
Polge, Chris. Interview with Roger Highfield, Royal Society, London, 23 June 1987.

Journal and Newspaper Articles
Aréchaga, J. 'Embryo culture, stem cells and experimental modification of embryonic genoma: An interview with Ralph Brinster'. *International Journal of Developmental Biology* 42, no. 7 (1998): 861–77.
Brinster, R. L. 'Germline stem cell transplantation and transgenesis'. *Science* 296 (21 June 2002): 2174–6.
Evans, M. J. and M. H. Kaufman. 'Establishment in culture of pluripotent cells from mouse embryos'. *Nature* 292 (9 July 1981): 154–6.
Gordon, Jon W. et al. 'Genetic transformation of mouse embryos by microinjection of purified DNA'. *Proceedings of the National Academies of Science* 77 (1980): 7380–4.
Hammer, R.E. et al. 'Production of transgenic rabbits, sheep and pigs by microinjection'. *Nature* 315 (20 June 1985): 680–3.

Highfield, Roger. 'Brave new era dawning down on the farm'. *Daily Telegraph*, 24 June 1987, p. 4.

Highfield, Roger. 'Chimeras help scientists in battle against genetic illness'. *Daily Telegraph*, 21 July 1988, p. 8.

Polge, C., S. Salamon and I. Wilmut. 'Fertilizing capacity of frozen boar semen following surgical insemination'. *Veterinary Record* 87, no. 15 (1970): 424.

Simons, Paul et al. 'Gene transfer into sheep'. *Bio/Technology* 6 (1988): 179.

Wilmut, I. and Lea Rowson. 'Successful low-temperature preservation of mouse and cow embryos'. *Journal of Reproduction and Fertility* 33, no. 2 (1973): 352–3.

Wilmut, I. and Lea Rowson. 'Experiments on low-temperature preservation of cow embryos'. *Veterinary Record* 92, no. 26 (1973): 686–90.

Book

Levick, Stephen. *Clone Being: Exploring the Psychological and Social Dimensions*. Lanham, Maryland: Rowman & Littlefield, 2004.

Report

Mori/Department of Trade and Industry. 'Science in society. Findings from qualitative and quantitative research'. London, March 2005

Internet Sources

Rosslyn Chapel Official Website <http://www.rosslynchapel.org.uk>.

Roslin Institute <http://www.roslin.ac.uk>.

Department of Trade and Industry homepage <http://www.dti.gov.uk>.

Chapter 3: A Brief History of Cloning

Interviews, Correspondence and Speeches

Bromhall, Derek. E-mail correspondence with Roger Highfield, 26 April and 11 May 2005.

Emlen, Stephen T. 'Sexual differences in mating strategies: The interface of culture law and biology'. Speech given to the annual meeting of the American Association for the Advancement of Science, Seattle, Washington, 14 February 1997.

Galli, Cesare. E-mail correspondence with Roger Highfield, 13 and 14 April 2005.

Graham, Chris. E-mail correspondence with Roger Highfield, 15 April 2005.

Graham, Chris. Phone interview with Roger Highfield, 15 April 2005.

Gurdon, John. E-mail correspondence with Roger Highfield, 14 and 22 March, 1 and 26 April, and 12 July 2005.

Haldane, J. B. L. 'Biological possibilities for the human species in the next ten thousand years'. Speech given in 1963, reprinted in Gordon Wolstenholme (editor) *Man and His Future*. Boston: Little, Brown, 1963.

Ritchie, Marjorie. Interview with author. Roslin Institute, 14 June 2005.

Schatten, Gerald. E-mail correspondence with Roger Highfield, 27 January, 6 March and 1 August 2005.

Schatten, Gerald. Phone interview with Roger Highfield, 17 May 2005.

Schatten, Gerald. Press conference, Royal Institution, London, 19 May 2005.

Solter, Davor. E-mail correspondence with Roger Highfield, 16, 18 and 30 March, 4, 7 and 19 April, 31 May and 4 June 2005.

Tarkowski, Andrzej. E-mail correspondence with Roger Highfield, 25 and 26 April and 6 May 2005.

Willadsen, Steen. E-mail correspondence with Roger Highfield, 12 October 2004.

Journal and Newspaper Articles

Briggs, R. and T. J. King. 'Transplantation of living nuclei from blastula cells into enucleated frogs eggs'. *Proceedings of the National Academy of Sciences* 38 (1952): 455–63.

Briggs, R. and T. J. King. 'Nuclear transplantation studies on the early gastrula (*Rana pipiens*). I. Nuclei of presumptive endoderm'. *Developmental Biology* 2 (1960): 252.

Bromhall, J. D. 'Nuclear transplantation in the rabbit egg'. *Nature* 258 (25 December 1975): 719–22.

Budiansky, S. 'Karl Illmensee: NIH withdraws research grant'. *Nature* 309 (19 April 1984): 738.

Campbell, K. H. S., J. McWhir, W. A. Ritchie and I. Wilmut. 'Sheep cloned by nuclear transfer from a cultured cell line'. *Nature* 380 (7 March 1996): 64–6.

Chan, A. W. S. et al. 'Clonal propagation of primate offspring by embryo splitting'. *Science* 287 (14 January 2000): 317–19.

Driesch, H. 'Zur Verlagerung der Blastomeren des Echinideneies'. *Anatomischer Anzeiger* 8 (1893): 348–57.

Fishberg, M., J. B. Gurdon and T. R. Elsdale. 'Sexually mature individuals of *Xenopus laevis* from the transplantation of single somatic nuclei'. *Nature* 182 (5 July 1958): 64–5.

Gurdon, J. B. N. and J. Byrne. 'The first half-century of nuclear transplantation'. *Proceedings of the National Academy of Sciences* 100 (2003): 8048–52.

Gurdon, J. B., R. A. Laskey and O. R. Reeves. 'The developmental capacity of nuclei transplanted from keratinized skin cells of adult frogs'. *Journal of Embryology and Experimental Morphology* 34 (1975): 93–112.

Harris H. and J. F. Watkins. 'Hybrid cells derived from mouse and man: Artificial heterokaryons of mammalian cells from different species'. *Nature* 205 (13 February 1965), 640–6.

Highfield, Roger. 'Cloned sheep have endless ramifications – Where sheep are sheep and men are uneasy'. *Daily Telegraph*, 7 March 1996, p. 1.

Highfield, Roger. 'Designer sperm "can weed out" gene defects'. *Daily Telegraph*, 23 October 2001, p. 13.

Hoppe, Peter C. and Karl Illmensee. 'Microsurgically produced homo-zygous-diploid uniparental mice'. *Proceedings of the National Academy of Sciences* 74 (1977): 5657–61.

Illmensee, K. and P. C. Hoppe. 'Nuclear transplantation in *Mus musculus*: Developmental potential of nuclei from preimplantation embryos'. *Cell* 23 (1981): 9–18.

McGrath, J. and D. Solter. 'Nuclear transplantation in the mouse embryo by microsurgery and cell fusion'. *Science* 220 (1983): 1300.

McGrath, J. and D. Solter. 'Inability of mouse blastomere nuclei transferred to enucleated zygotes to support development *in vitro*'. *Science* 226 (1984): 1317–19.

McLaren, Anne. 'Cloning: Pathways to a pluripotent future'. *Science* 288 (9 June 2000): 1775–80.

Marx, J. L. 'Bar Harbor investigation reveals no fraud'. *Science* 220 (1983): 1254.

Newmark, Peter. 'Geneva and Bar Harbor Labs to check results'. *Nature* 303 (2 June 1983): 363.

Solter, D. 'Imprinting'. *International Journal of Developmental Biology* 42 (1998): 951–4.

Solter, D. 'Mammalian cloning: Advances and limitations'. *Nature Reviews Genetics* 3 (2000):199–207.

Tarkowski, A. K. 'Mouse chimaeras revisited: Recollections and reflections'. *International Journal of Developmental Biology* 42, no. 7 (1998): 903–8.

Tsunoda, Y. and Y. Kato. 'Full-term development after transfer of nuclei from 4-cell and compacted morula stage embryos to enucleated oocytes in the mouse'. *Journal of Experimental Zoology* 278 (1997): 250–4.

Wakayama, T. et al. 'Mice cloned from embryonic stem cell'. *Proceedings of the National Academy of Sciences* 96 (1999): 14984–9.

Willadsen, S. M. 'Nuclear transplantation in sheep embryos'. *Nature* 320 (6 March 1986): 63–5.

Woods, G.L. et al. 'A mule cloned from fetal cells by nuclear transfer'. *Science* 301 (22 August 2003): 1063.

Books

Cibelli, J., R. P. Lanza, H. S. Campbell and Michael D. West. *Principles of Cloning*. San Diego, California: Elsevier, 2002.

Huxley, Aldous. *Brave New World*. New York: Harper, 1998.

Mitchison, N. *Solution Three*. New York: The Feminist Press, 1995.

Rorvik, David. *In His Image: The Cloning of a Man*. Philadelphia: Lippincott, 1978.

Spemann, H. *Embryonic Development and Induction*. New Haven, Connecticut: Yale University Press, 1938.

Toffler, Alvin. *Future Shock*. New York: Random House, 1970.

Weismann, A. *The Germ-Plasm Theory of Heredity*. New York: Charles Scribner's Sons, 1902.

Wilmut, I., K. Campbell and C. Tudge. *The Second Creation*. London: Headline, 2000.

Internet Sources

The Nobel Foundation <http://www.nobelprize.org>.

Brownlee, Christen. 'Nuclear transfer: Bringing in the clones'. *Proceedings of the National Academy of Science* <http://www.pnas.org/misc/classics4.shtml>.

Genetics Savings & Clone, Inc. <http://www.savingsandclone.com>.

Television Programme

Brown, Louise. *The Baby Makers*, London, Channel 4, 1999.

Chapter 4: Dollymania

Interviews and Correspondence

Archibald, Alan. Interview with Roger Highfield, Roslin Institute, 26 September 2003.

Archibald, Alan. E-mail correspondence with Roger Highfield, 10 February 2005.

Bracken, John. Interview with Roger Highfield, Roslin Institute, 14 June 2005.

Campbell, Keith. Phone interview with Roger Highfield, 16 June 2005.

Campbell, Keith. E-mail correspondence with Roger Highfield, 4 and 5 July 2005.

Colman, Alan. E-mail correspondence with Roger Highfield, 14 and 25 April 2005.

Larsson, Nils-Göran. E-mail correspondence with Roger Highfield, 26 May 2004.

Ritchie, Marjorie. Interview with Roger Highfield, Roslin Institute, 14 June 2005.

Walker, Karen (née Mycock). E-mail correspondence with Roger Highfield, 7, 8, 12 and 24 July 2005.

Westhusin, Mark. E-mail correspondence with Roger Highfield, 9 February 2004, 2 August 2005.

Yang, Jerry. E-mail correspondence with Roger Highfield, 14 April and 23 June 2005.

Journal and Newspaper Articles

Ashworth, D. et al. 'DNA microsatellite analysis of Dolly'. *Nature* 394 (23 July 1998): 329.

Campbell, K. H. S., P. Loi, P. J. Otaegui and I. Wilmut. 'Cell cycle co-ordination in embryo cloning by nuclear transfer'. *Reviews of Reproduction* 1 (1996): 40–6.

Galli, C. et al. 'A cloned horse born to its dam twin'. *Nature* 424 (7 August 2003): 635.

Highfield, Roger. 'She looks exactly like any other sheep. But the cloning method used to produce Dolly may change our lives'. *Daily Telegraph*, 24 February 1997, p. 5.

Mckie, Robin. 'Scientists clone adult sheep'. *Observer*, 23 February 1997, p. 1.

Pell, Cardinal George. 'Scientific spin creates a sub class of humans'. *Sunday Telegraph (Sydney)*, 26 June 2005, p. 99.

Signer, E. et al. 'DNA fingerprinting Dolly'. *Nature* 394 (23 July 1998): 329–30.

Solter, D. 'Lambing by nuclear transfer'. *Nature* 380 (7 March 1996): 24.

Solter, D. 'Dolly is a clone – and no longer alone'. *Nature* 394 (23 July 1998): 315–16.

Thompson, J. G. et al. 'Lamb birth weight is affected by culture system utilized during *in vitro* pre-elongation development of ovine embryos'. *Biology of Reproduction* 53 (1 December 1995): 1385–91.

Tian, Cindy et al. 'Meat and milk compositions of bovine clones'. *Proceedings of the National Academy of Sciences* 102 (3 May 2005): 6261–6.

Vittorio, Sgaramella and Norton D. Zinder. 'Letters'. *Science* 279 (30 January 1998): 635–6.

Wade, Nicholas. 'Scientists who announced cloning of sheep Dolly is taking steps to win over critics'. *New York Times*, 28 February 1998, p. A6.

Wakayama, T. et al. 'Full-term development of mice from enucleated oocytes injected with cumulus cell nuclei'. *Nature* 394 (23 July 1998): 369–74.

Wilmut, I. et al. 'Viable offspring derived from foetal and adult mammalian cells'. *Nature* 385 (27 February 1997): 810–13.

Walker, S. K., J. L. Hill, D. O. Kleemann and C. D. Nancarrow. 'Development of ovine embryos in synthetic oviductal fluid containing amino acids at oviductal fluid concentrations'. *Biology of Reproduction* 55 (1 September 1996): 703–8.

Wilson, Frederick H. et al. 'A cluster of metabolic defects caused by mutation in a mitochondrial tRNA'. *Science* 306 (12 November 2004): 1190–4.

Books

Kolata, Gina. *Clone: The Road to Dolly and the Path Ahead*. London: Allen Lane, 1997.

Wilmut, I., K. Campbell and C. Tudge. *The Second Creation*. London: Headline, 2000.

Report

Nuffield Foundation for Bioethics. 'The ethics of research involving animals'. 25 May 2005.

Internet Sources

Advanced Cell Technology <http://www.advancedcell.com>.

Alexion Pharmaceuticals Inc. <http://www.alexionpharmaceuticals. com>.

Berg, Paul. 'Asilomar and recombinant DNA'. The Nobel Foundation <http://nobelprize.org/medicine/articles/berg/index.html >.

Biotechnology and Biological Sciences Research Council <http://www.bbsrc.ac.uk>.

Medarex <http://www.medarex.com>.

PharmingGroup, NV <http://www.pharming.com>.

Roman Cuira Pontifical Academies <http://www.vatican.va/ roman_curia/pontifical_academies/acdscien/>.

Foundation on Economic Trends Homepage <http://www.foet.org>.

Royal Society <http://www.royalsoc.ac.uk>.

Society, Religion and Technology Project <http://www.srtp.org.uk>.

Uncaged Campaigns: Xenotransplantation <http://www.uncaged.co.uk/ xeno.htm>.

Chapter 5: Farmyard Clones

Correspondence

Christmas, Robin. E-mail correspondence with Roger Highfield, 10 and 12 July 2005.

Colman, Alan. E-mail correspondence with Roger Highfield, 14, 17 and 25 April 2005.

Robl, Jim. E-mail correspondence with Roger Highfield, 23 and 24 February, 8 and 13 April 2005.

Rosenbaum, Linda. Email correspondence with Roger Highfield, 8 July 2005.

Tuddenham, Ted. E-mail correspondence with Roger Highfield, 17 March, 4, 5, 8 and 11 July 2005.

Journal and Newspaper Articles

Briggs, R. et al. 'Christmas disease: A condition previously mistaken for haemophilia'. *British Medical Journal* 2 (27 December 1952): 1378–82.

Clark, A. J. et al. 'Expression of human anti-hemophilic factor IX in the milk of transgenic sheep'. *Bio/Technology* 7 (1989): 487–92.

Clark, John and Bruce Whitelaw. 'A future for transgenic livestock'. *Nature Reviews Genetics* 4 10 (2003): 825–33.

Dai, Y. et al. 'Targeted disruption of the alpha1,3-galactosyltransferase gene in cloned pigs'. *Nature Biotechnology* 20, no. 3 (2002): 251–5.

Fire, A. et al. 'Potent and specific genetic interference by double-stranded RNA in *Caenorhabditis elegans*'. *Nature* 391 (19 February 1998): 806–11.

Giangrande, Paul. 'Six characters in search of an author: The history of the nomenclature of coagulation factors'. *British Journal of Haematology* 121 (2003): 703–12.

Hamilton, A. J. and D. C. Baulcombe. 'A novel species of small antisense RNA in post-transcriptional gene silencing'. *Science* 286 (29 October 1999): 950–2.

Highfield, Roger. 'Brave new world for man and beast down on the farm'. *Daily Telegraph*, 6 July 1987, p. 9.

Highfield, Roger. 'Farm animals "chemical factories" of the future'. *Daily Telegraph*, 11 August 1987, p. 12.

Lai, L. X. et al. 'Production of alpha-1,3-galactosyltransferase knockout pigs by nuclear transfer cloning'. *Science* 295 (8 February 2002): 1089–92.

McCreath, K. et al. 'Production of gene-targeted sheep by nuclear transfer from cultured somatic cells'. *Nature* 405 (29 June 2000): 1066–9.

Mannucci, P. M. and E. G. Tuddenham. 'The hemophilias – from royal genes to gene therapy'. *New England Journal of Medicine* 344 (7 June 2001): 1773–9.

Schnieke, A. E. et al. 'Human factor IX transgenic sheep produced by transfer of nuclei from transfected fetal fibroblasts'. *Science* 278 (19 December 1997): 2130–3.

Wall, Robert J. et al. 'Genetically enhanced cows resist intramammary *Staphylococcus aureus* infection'. *Nature Biotechnology* 23 (1 April 2005): 445–51.

Weiss, Robin. 'Xenografts and retroviruses'. *Science* 285 (20 August 1999): 1221–2.

Reports

Nuffield Council on Bioethics. 'Animal-to-human transplants: The ethics of xenotransplantation'. London, 1996.

Institute of Medicine. 'Xenotransplantation, science, ethics, and public policy'. Washington, DC, 1996.

Department of Health Advisory Group. 'Department of Health Advisory Group on the Ethics of Xenotransplantation Animal Tissues into Humans'. London: Her Majesty's Stationary Office, 1997.

Internet Sources

Canadian Hemophilia Society <http://www.hemophilia.ca>.

Geron <http://www.geron.com>.

GTC Biotherapeutics <http://www.transgenics.com>.

Hematech <http://www.hematech.com>.

Nexia Biotechnologies, Inc. <http://www.nexiabiotech.com>.

Chapter 6: Cloning for My Father

Interviews, Correspondence and Briefings

Andrews, Peter. E-mail correspondence with Roger Highfield, 9 and 10 February 2004, 25 June 2005.

Blakemore, Colin. Interview with Roger Highfield, UK Stem Cell Bank, South Mimms, 19 May 2004, 18 January 2005, London.

Cibelli, Jose. E-mail correspondence with Roger Highfield, 20 January 2006.

De Sousa, Paul. E-mail correspondence with Roger Highfield, 4 June 2004, 5 and 7 March 2005.

De Sousa, Paul. British Association for the Advancement of Science, Dublin, 9 September 2005, and Roslin Institute press briefing, 3 March 2005.

Hwang, W. S. E-mail correspondence with Roger Highfield, 10 February 2004; 21 May, 29 July, 2 August and 25 Novermber 2005.

Hwang, W. S. Press conferences at the American Association for the Advancement of Science. Seattle, Washington, 12 February 2004.

Hwang, W. S. Press conference with Gerald Schatten at the Royal Institution, London, 19 May 2005.

Lovell-Badge, Robin. Interview with Roger Highfield. Science Media Centre, London, 31 May 2005.

McWhir, Jim and Wei Cui. Roslin Institute press briefing, 3 March 2005.

Minger, Stephen. Interviews with Roger Highfield. London, 12 January 2005.

Minger, Stephen. E-mail correspondence with Roger Highfield, 13 January and 25 May 2005.

Murdoch, Alison. E-mail correspondence with Roger Highfield. 13 and 25 May, 15 and 24 June, 18 and 27 August 2004; 1 February and 19 May 2005.

Murdoch, Alison. Interviews with Roger Highfield. Centre for Life, Newcastle upon Tyne, 19 October 2004.

Pedersen, R. 'Stem cells 2005: Progress to therapy'. Talk given at the Edinburgh International Conference Centre, 4–5 March 2005.

Pedersen, Roger. Interview with Roger Highfield, Science Media Centre, London, 31 May 2005.

Schatten, Gerald. E-mail correspondence with Roger Highfield, 8 January 2006.

Shaw, Christopher. Interview with Roger Highfield. Science Media Centre, London, 28 September 2004.

Smith, Austin. E-mail correspondence with Roger Highfield, 23 November 2004; 19 January, 3 March, 27 June, and 15, 16, 22, 23 and 25 August 2005.

Smith, Austin. Interview with Roger Highfield, The Tun, Edinburgh, March 3, 2005

Stojkovic, Miodrag. Interview with Roger Highfield. Centre for Life, Newcastle upon Tyne, 19 October 2004.

Stojkovic, Miodrag. E-mail correspondence with Roger Highfield, 15 and 24 June, 28 July, 22 September, 13 October, 22 November and 22 December 2004; 13, 18 and 19 April, and 19 May 2005; and 2, 9 and 10 January 2006.

Journal and Newspaper Articles

Faiola, Anthony and Rick Weiss. 'South Korean panel debunks scientist's stem cell claims; fraud finding is another setback in cloned embryo research'. *Washington Post Foreign Service*, 10 January 2006.

Highfield, Roger. 'Finding the wonder cell'. *Daily Telegraph*, 25 July 1988, p. 8.

Highfield, Roger and David Derbyshire. 'Human cells cloned: Babies next?' *Daily Telegraph*, 13 February 2004, p. 1.

Highfield, Roger. 'Scientists take a giant step forward in human cloning'. *Daily Telegraph*, 20 May 2005, p. 1.

Highfield, Roger. 'Have we been oversold the stem cell dream?' *Daily Telegraph*, 29 June 2005, p. 21.

Highfield, Roger. 'New doubts cast on human clone claim'. *Daily Telegraph*, 15 December 2005, p. 10.

Highfield, Roger. 'Cloning star who fooled the world'. *Daily Telegraph*, 24 December 2005, p. 4.

Highfield, Roger. 'Embryo cloning cheat resigns in disgrace'. *Daily Telegraph*, 24 December 2005, p. 1.

Highfield, Roger. 'Disgraced cloning expert blames colleague'. *Daily Telegraph*, 27 December 2005, p. 2.

Highfield, Roger. 'Clone doctor faked his work on stem cells condemned by panel'. *Daily Telegraph*, 11 January 2006, p. 14.

Highfield, Roger. 'Disgraced doctor "blinded by zeal"'. *Daily Telegraph*, 13 January 2006, p. 10.

Hwang, Woo Suk et al. 'Evidence of a pluripotent human embryonic stem cell line derived from a cloned blastocyst'. *Science* 303 (12 March 2004): 1669–74.

Hwang, Woo Suk et al. 'Patient-specific embryonic stem cells derived from human SCNT blastocysts'. *Science* 308 (17 June 2005): 1777–83.

Jiang, Y. et al. 'Pluripotent nature of adult marrow derived mesenchymal stem cells'. *Nature* 418 (4 July 2002): 41–9.

Kono, Tomohiro et al. 'Birth of parthenogenetic mice that can develop to adulthood'. *Nature* 428 (22 April 2004): 860–4.

Marshall, Eliot and Gretchen Vogel. 'Cloning announcement sparks debate and scientific skepticism' *Science* 294 (30 November 2001): 1802–3.

Martin, Maria, Alysson Muotri, Fred Gage and Ajit Varki. 'Human embryonic stem cells express an immunogenic nonhuman sialic acid'. *Nature Medicine* 11 (2005): 228–32.

Park, Alice and Stella Kim. 'Independent panel confirms stem cell fraud. Outside investigators determine that 2004 cloned stem cell lines from South Korean lab were faked'. *Time Online*, 9 January 2006.

Pickering, Susan J. et al. 'Generation of a human embryonic stem cell line encoding the cystic fibrosis mutation Δf508, using preimplantation genetic diagnosis'. *Reproductive BioMedicine Online* 10 (March 2005): 390–7.

Samuel Reich, Eugenie. 'Stem-cell pioneer's findings in doubt; Woo Suk Hwang is in trouble again, this time over his claims to have made personalised cloned cells'. *New Scientist*, 17 December 2005, Vol. 188; Issue 2530, p. 6.

Shamblott, M. J. et al. 'Derivation of pluripotent stem cells from cultured human primordial germ cells'. *Proceedings of the National Academy of Sciences* 95 (1998): 13726–31

Simerly, Calvin et al. 'Molecular correlates of primate nuclear transfer fail-
    ures'. *Science* 300 (11 April 2003): 297.
Skloot, R. 'Henrietta's dance'. *Johns Hopkins Magazine* (April 2000, online
    edition).
Stojkovic, Miodrag et al. 'Derivation of a human blastocyst after heterol-
    ogous nuclear transfer to donated oocytes' *RBM Online* (2005) Volume
    11, No. 2: 226–31.
Taylor, C. J. et al. 'Banking on human embryonic stem cells: Estimating the
    number of donor cell lines needed for HLA matching'. *The Lancet* 366
    (10 December 2005): 2019–25.
Thomson, James A. et al. 'Embryonic stem cell lines derived from human
    blastocysts'. *Science* 282 (6 November 1998): 1145–7.
Ying, Q. L., J. Nichols, E. P. Evans and A. G. Smith. 'Changing potency by
    spontaneous fusion'. *Nature* 416 (4 April 2002): 545–8.

Reports
Chief Medical Officer's Expert Group on Therapeutic Cloning. 'Stem cell
    research: Medical progress with responsibility'. United Kingdom
    Department of Health, 2000.
House of Lords Select Committee. 'Stem cell research'. London,
    27 February 2002.
Human Fertilisation & Embryology Authority/Human Genetics Advisory
    Committee. 'Cloning issues in reproduction, science and medicine'
    London, 1998.
Royal Society. 'Whither cloning?'. London, 1 January 1998.
Summary of the Final Report on Professor Woo Suk Hwang's Research
    Allegations, Seoul National University Investigation Committee,
    9 January 2006.
Warnock, Dame Mary. 'Report of the Committee of Inquiry into Human
    Fertilisation and Embryology'. London: Her Majesty's Stationary
    Office, 1984.

Internet Sources
Advanced Cell Technology <http://www.advancedcell.com>.
Centre for Life <http://www.nfc-life.org.uk>.
Comment on Reproductive Ethics <http://www.corethics.org>.
Human Fertilisation & Embryology Authority <http://www.hfea.
    gov.uk>.
Human Genetics Commission <http://www.hgc.gov.uk>.
Nuffield Council on Bioethics <http://www.nuffield.org.uk/ bioethics/>.
Royal Society: Science, Policy and Government <http://www.royal-
    soc.ac.uk/policy/>.

UK Stem Cell Bank <http://www.ukstemcellbank.co.uk>.
Other stem cell research centres and sources of information include
    <http://www.iscr.ed.ac.uk>, <http://www.stemcells.cam.ac.uk>,
    <http://www.snu.ac.kr>, <http://www.stemcell.harvard.edu>,
    <http://stemcells.nih.gov>, <http://www.nscc.edu.au>,
    <http://www.stemcellforum.org>, <http://www.isscr.org>.

Unpublished Material
Hwang, W. S. Notes handed to journalists at the meeting of the American
    Association for the Advancement of Science, 12 February 2004.

Chapter 7: Is a Blastocyst a Person?

Interviews, Correspondence and Press Material
Banner, Michael. E-mail correspondence with Roger Highfield,
    24 March, 7 April 2005.
Banner, Michael. Phone interview with Roger Highfield, 18 March 2005.
Cibelli, Jose. E-mail correspondence with Roger Highfield, 7 January 2005.
Edwards, Robert. Press conference, Queen Elizabeth II Conference
    Centre, London, Sixth PGD Symposium, 19 May 2005.
Harris, John. Interview with Roger Highfield, Eurostem meeting,
    Manchester, 24 and 25 February 2005.
Harris, John. E-mail correspondence with Roger Highfield, 15 and 29
    March, 16 and 17 May 2005.
Life. 'Another step down the road to trivialization of human life'. Press
    release, 27 February 2002.
Zernicka-Goetz, Magda. E-mail correspondence with Roger Highfield,
    20 May, 10 and 22 June 2005.

Journal and Newspaper Articles
Allen, Mike. 'Destruction of life is part of culture of death'. *Lexington
    Herald-Leader*, 6 September 2005, p. a6.
Cleland, Carol E. and Christopher F. Chyba. 'Defining life'. *Origins of Life
    and Evolution of the Biosphere* 32, no. 4 (2002): 387–93.
Connor, Kenneth. 'Guard sanctity of life'. *USA Today*, 10 August 2001,
    p. 10A.
Cookson, Clive and Daniel Green. 'A wolf in sheep's clothing?'. *Financial
    Times*, 6 December 1997, p. 6.
Editorial. 'Is it wrong to clone humans?' The *Express*, 13 February 2004,
    p. 12.
Editorial. 'Bush advocates smartest position on stem cells'. *Buffalo News*,
    15 August 2004, p. H5.

Greene, Mark et al. 'Moral issues of human–non–human primate neural grafting'. *Science* 309 (15 July 2005): 385–6.

Harper, J. C. and J. D. A. Delhanty. 'Preimplantation genetic diagnosis'. *Current Opinions in Obstetrics and Gynaecology* 12 (April 2000): 67–72.

Steptoe, P. and R. Edwards. 'Birth after the reimplantation of a human embryo'. *The Lancet* 2, no. 8085 (12 August 1978): 366.

Timmons, Heather. 'UK clears cloning of human embryos'. *International Herald Tribune*, 12 August 2004, p. 1.

Books

Barrow, John. *Pi in the Sky*. Oxford: Clarendon Press, 1992.

Coveney, Peter and Roger Highfield. *Frontiers of Complexity*. London: Faber, 1995.

Harris, John. *The Value of Life*. London: Routledge, 1985.

Klotzko, Arlene. *A Clone of Your Own? The Science and Ethics of Cloning*. Oxford: Oxford University Press, 2004.

Schrödinger, Erwin. *What Is Life?* with *Mind and Matter* and *Autobiographical Sketches*. Cambridge: Canto, 1992.

Reports

Human Genetics Advisory Commission and Human Fertilisation & Embryology Authority. 'Cloning issues in reproduction, science and medicine'. London, January 1998.

Medical Research Council (MRC). 'Report of the MRC Expert Group on Fetal Pain'. London, 28 August 2001.

Internet Sources

Comment on Reproductive Ethics <http://www.corethics.org>.

Eurostem <http://www.eirma.asso.fr/eurostem.html>.

Life ('The UK's Leading Pro-Life Charity') <http://www.lifeuk.org>.

Motor Neurone Disease Association <http://www.mndassociation.org>.

National Right to Life <http://www.nrlc.org>.

Nuffield Council on Bioethics <http://www.nuffieldbioethics.org>.

President George W. Bush <http://www.whitehouse.gov>.

Select Committee on Science and Technology <http://www.parliament.uk>.

Chapter 8: Why We Should Not Clone Babies

Correspondence

Collas, Philippe. E-mail correspondence with Roger Highfield, 5 and 6 January 2005.

Ding, Sheng. E-mail correspondence with Roger Highfield, 7 January 2005.

Krawetz, Stephen. E-mail correspondence with Roger Highfield, 5 and 7 April, 2 August 2005.

Lai, Tony. E-mail correspondence with Roger Highfield, 13 November 2002.

Lanza, Robert. E-mail correspondence with Roger Highfield, 23 September 2004, 5 January and 29 June 2005.

Renard, Jean-Paul. E-mail correspondence with Roger Highfield, 6 April 2005.

Skok, Jane. E-mail correspondence with Roger Highfield, 10, 11, 14 and 15 March 2005.

Young, Lorraine. E-mail correspondence with Roger Highfield. 18 October 2004.

Journal and Newspaper Articles

Dawkins, Richard. 'The secret confessions of a closet clone'. *Evening Standard*, 25 February 1997, p. 9.

De Robertis, E. M., G. A. Partington, R. F. Longthorne and J. B. Gurdon. 'Somatic nuclei in amphibian oocytes: Evidence for selective gene expression'. *Journal of Embryology and Experimental Morphology* 40 (1977): 199–214.

Heyman Y. et al. 'Frequency and occurrence of late-gestation losses from cattle cloned embryos'. *Biology of Reproduction* 66, no. 1 (January 2002): 6–13.

Highfield, Roger. 'Human eggs can be mass produced from male embryos'. *Daily Telegraph*, 2 May 2003, p. 11.

Highfield, Roger. 'Scientist predicts end of infertility within 10 years'. *Daily Telegraph*, 25 July 2003, p. 6.

Highfield, Roger. 'The Holy Grail of stem-cell research' *Daily Telegraph*, 19 January 2005, p. 16.

Highfield, Roger. 'Check on gene therapy after leukemia scare in France'. *Daily Telegraph*, 4 March 2005, p. 11.

Krawetz, Stephen. 'Paternal contribution: New insights and future challenges'. *Nature Reviews Genetics* 6 (2005): 633–42.

Lanza, Robert et al. 'Extension of cell life-span and telomere length in animals cloned from senescent somatic cells'. *Science* 288 (28 April 2000): 665–9.

Peura, T. 'Improved *in vitro* development rates of sheep somatic nuclear transfer embryos by using a reverse-order zona-free cloning method'. *Cloning and Stem Cells* 5, no. 1 (March 2003): 13–24.

Ribas, R. et al. 'Development of a zona-free method of nuclear transfer in the mouse'. *Cloning and Stem Cells* 7, no. 2 (2005): 126–38.

Rideout, William M., Kevin Eggan and Rudolf Jaenisch. 'Nuclear cloning and epigenetic reprogramming of the genome'. *Science* 293 (10 August 2001): 1093–98.

Sakatani, Takashi et al. 'Loss of imprinting of *Igf2* alters intestinal maturation and tumorigenesis in mice'. *Science* 307 (25 March 2005): 1976–8.

Wakayama, T. et al. 'Cloning of mice to six generations'. *Nature* 407 (21 September 2000): 318–19.

Wells, D. N. B. Oback and G. Laible. 'Cloning livestock: A return to embryonic cells'. *Trends in Biotechnology* 21 (2003): 428–32.

Wells, D. N., J. T. Forsyth, V. McMillan and B. Oback. 'The health of somatic cell cloned cattle and their offspring'. *Cloning Stem Cells* 6, no. 2 (2004):101–10.

Zhou, O. et al. 'Generation of fertile cloned rats by regulating oocyte activation'. *Science* 302 (14 November 2003): 1179.

Books

Levick, Stephen E. *Clone Being: Exploring the Psychological and Social Dimensions.* Lanham, Maryland: Rowman & Littlefield, 2004.

Internet sources

AgResearch <http://www.agresearch.co.nz>.

British Fertility Society <http://www.britishfertilitysociety.org.uk>.

National Institute for Agricultural Research (INRA) <http://www.inra.fr/english/>.

Television Programme

Brown, Louise. *The Baby Makers*, London, Channel 4, 1999.

Chapter 9: Designer Babies, and Envoi: Beyond Human Cloning

Interviews, Correspondence and Press Conferences

Braude, Peter. Interview with Roger Highfield. Science Media Centre, London, 25 April 2005.

Delhanty, Joy. E-mail correspondence with Roger Highfield, 6, 14, 22 and 25 January, and 31 March 2005.

Fischer, Alain. E-mail correspondence with Roger Highfield, 26 April 2000, 9 October 2004, 2 March 2005.

Gaspar, Bobby. Interview with Roger Highfield. Great Ormond Street Hospital, London, 26 June 2003.

Handyside, Alan. Press conference, London, 18 April 1990.

Thrasher, Adrian. Interview with Roger Highfield, Great Ormond Street Hospital, London, 26 June 2003.

Thrasher, Adrian. E-mail correspondence with Roger Highfield, 7 October and 1 November 2004; 25 February, 1, 3, 9, 11 and 30 March, 22 April and 30 June 2005.

Turnbull, Doug. E-mail correspondence with Roger Highfield. 15 March, 15 and 30 August, and 2, 4, 7 and 8 September 2005.

Watson, James. Press conference. Queen Elizabeth II Conference Centre, London, sixth PGD Symposium, 19 May 2005; Press conference, Royal Society, 23 April 2003.

Winston, Robert. Interviews with Roger Highfield, London, 18 April 1990; Royal Society, London, 26 January 2000; Dublin, 9 September 2005.

Journal and Newspaper Articles

Barritt, J., S. Willadsen, C. Brenner and J. Cohen. 'Cytoplasmic transfer in assisted reproduction'. *Human Reproduction* 7, no. 4 (2001): 428–35.

Boyle, Robert and Julian Savulescu. 'Ethics of using preimplantation genetic diagnosis to select a stem cell donor for an existing person'. *British Medical Journal* 323 (24 November 2001): 1240–3.

Cavazzana-Calvo, Marina et al. 'Gene therapy of human severe combined immunodeficiency (SCID)-X1 disease'. *Science* 288 (28 April 2000): 669–72.

Chan, A. W. S. et al. 'Transgenic monkeys produced by retroviral gene transfer into mature oocytes'. *Science* 291 (12 January 2001): 309–12.

Hacein-Bey-Abina, S. et al. '*LMO2*-associated clonal T cell proliferation in two patients after gene therapy for SCID-X1'. *Science* 302 (17 October 2003): 415–19.

Handyside, A. H., E. H. Kontogianni, K. Hardy and R. M. L. Winston. 'Pregnancies from biopsied human preimplantation embryos sexed by Y-specific DNA amplification'. *Nature* 344 (19 April 1990): 768–70.

Harris, J. 'Intimations of immortality'. *Science* 288 (7 April 2000): 59.

Highfield, Roger. 'The beginning of the end for fatal hereditary disease'. *Daily Telegraph*, 18 January 1988, p. 12.

Highfield, Roger. 'Scientists call for limits to "designer baby" transplants'. *Daily Telegraph*, 3 June 1988, p. 7.

Highfield, Roger. 'The ethics of having a baby as a spare part'. *Daily Telegraph*, 9 April 1990, p. 16.

Highfield, Roger. 'Researchers open the way to screening test-tube babies for genetic diseases'. *Daily Telegraph*, 19 April 1990, p. 1.

Highfield, Roger. 'Why life will never be the same again. The new genetics'. *Telegraph Magazine*, 17 August 1991, p. 26.

Highfield, Roger. 'How do you define a human being?' *Daily Telegraph*, 28 October 1991, p. 12.

Highfield, Roger. 'Meet the baby with three parents'. *Daily Telegraph*, 30 May 2001, p. 25.

Highfield, Roger. 'British team discovers gene lifeline for children'. *Daily Telegraph*, 28 June 2003, p. 15.

Highfield, Roger. 'Is there a case for modifying her genes?' *Daily Telegraph*, 31 August 2005, p. 14.

Slater, Eve. 'Today's FDA'. *New England Journal of Medicine* 352, no. 3 (2005): 293.

Sturm, R. A. and T. Frudakis. 'Eye colour: Portals into pigmentation genes and ancestry'. *Trends in Genetics* 20 (2004): 327–32.

Thrasher, A. J. 'Gene therapy: Great expectations?' *Medical Journal of Australia*, 182 (2005): 440–1.

Wells, D. and J. Delhanty. 'Comprehensive chromosomal analysis of human preimplantation embryos using whole genome amplification and single cell comparative genomic hybridization'. *Molecular Human Reproduction* 6, no. 11 (2000): 1055–62.

Williams, Thomas et al. 'An immune basis for malaria protection by the sickle cell trait'. *Public Library of Science Medicine* 2, no. 5 (2005): 128

Books

Andrews, Lori. *The Clone Age*. New York: Owl Books, 2000.

Frankel, Mark and Audrey Chapman. *Human Inheritable Genetic Modifications*. Washington, DC: American Association for the Advancement of Science, 2000.

Klotzko, Arlene. *A Clone of Your Own? The Science and Ethics of Cloning*. Oxford: Oxford University Press, 2000.

Levick, Stephen. *Clone Being: Exploring the Psychological and Social Dimensions*. Lanham, Maryland: Rowman & Littlefield, 2004.

Silver, Lee. *Remaking Eden: Cloning and Beyond in a Brave New World*. New York: William Morrow, 1997.

Watson, James. *DNA: The Secret of Life*. London: Heinemann, 2003.

Internet sources

American Society for Reproductive Medicine <http://www.asrm.org>.

American Society of Gene Therapy <http://www.asgt.org>.

European Society of Human Reproduction and Embryology <http://www.eshre.com>.

The Huntington's Disease Association <http://www.hda.org.uk>.

Sickle Cell Society <http://www.sicklecellsociety.org>.

Index